HOW TO BECOME
A NAVY SEAL

HOW TO BECOME A NAVY SEAL

EVERYTHING YOU NEED TO KNOW TO BECOME A MEMBER OF THE U.S. NAVY'S ELITE FORCE

Don Mann

Skyhorse Publishing

Skyhorse Publishing books may be purchased in bulk at special discounts for sales promotion, corporate gifts, fund-raising, or educational purposes. Special editions can also be created to specifications. For details, contact the Special Sales Department, Skyhorse Publishing, 307 West

36th Street, 11th Floor, New York, NY 10018 or info@skyhorsepublishing.com.

Skyhorse® and Skyhorse Publishing® are registered trademarks of Skyhorse Publishing, Inc.®, a Delaware corporation.

Visit our website at www.skyhorsepublishing.com.

10 9 8 7 6 5 4 3 2 1

Library of Congress Cataloging-in-Publication Data
Mann, Don, 1957–
 How to become a Navy SEAL : everything you need to know to become a member of the U.S. Navy's elite force / Don Mann.
     pages cm
 ISBN 978-1-62087-186-7 (alk. paper)
 1. United States. Navy. SEALs. 2. United States. Navy. SEALs--Vocational guidance. I. Title.
 VG87.M25 2014
 359.9'84--dc23
                          2014015208

Cover design by Rain Saukas

Ebook ISBN: 978-1-62873-487-4

Printed in China

CONTENTS

INTRODUCTION

The Beginning

For outstanding performance in combat during the invasion of Normandy, June 6, 1944. Determined and zealous in the fulfillment of an extremely dangerous hazardous mission, the Navy Combat Demolition Unit of Force "O" landed on the "Omaha Beach" with the first wave under devastating enemy artillery machine-gun and sniper fire. With practically all explosives lost and with their force seriously depleted by heavy casualties the remaining officers and men carried on gallantly, salvaging explosives as they were swept ashore and in some instances commandeering bulldozers to remove obstacles. In spite of these grave handicaps, the Demolition crews succeeded initially in blasting five gaps through enemy obstacles for the passage of assault forces to the Normandy shore and within two days had sapped over eighty-five percent of the "Omaha Beach" area of German-placed traps. Valiant in the face of grave danger and persistently aggressive against fierce resistance, the Navy Combat Demolition Unit rendered daring and self-sacrificing service in the performance of a vital mission, thereby sustaining the high traditions of the United States Naval Service —Presidential Unit Citation, one of three awarded to the Navy for the Normandy landings.

In the fall of 1942, a small detachment of sailors—"demolitioneers"—having been given a crash course in cable cutting, commando attacks, and, of course, demolition at an amphibious training base in Little Creek, Virginia, sailed to French North Africa to conduct its first World War II mission: part of Operation Torch, a joint British-American invasion that, if successful, would give the Allies open access to the soft underbelly of Europe—Sicily and Italy—and ultimately Germany.

Under cover of darkness, the maritime commandos came to the mouth of the Wadi Sebou River in Morocco, 75 mm artillery fire raining down on them from the fortress of the Kasbah. Their mission was to remove the boom and net, held up by a heavy cable, that were blocking the port and the way to Port Lyautey Airfield, occupied by the Axis-friendly Vichy French military.

The operation was perilous—heavy squalls, enemy fire, and monster surf threatened to quash the mission of the 17 men in their Higgins boat, a shallow barge-like landing craft. The men reached the net, cut it with explosive cable cutters and raced downstream, gunfire from the Kasbah in their wake. Not one man was hit. The net was swept away, and the USS *Dallas* rammed through the river boom, unloading its troops, who captured the airfield.

These men, who were part one of the first successful Naval demolition teams, went back to the United States to help the military create more obstacle clearance units. By 1943 orders went out to form special combat demolition units, and swiftly. The enemy—Germany and the other Axis powers—would certainly create every manner of obstacle to slow the landing by sea of Allied troops in Axis strongholds such as Sicily, and the coast of Normandy in France. If forced to debark in heavy surf Allied troops could lose their weapons and worse, drown or be shot as they slogged through hip-high water.

⌃ **Troops wade ashore at Omaha Beach on D-Day, June 6, 1944.**

The Allied amphibious forces needed the support of advance demolition teams. These units would be sent first to the beaches of Sicily and a year later, to Normandy to open channels through the heavily fortified shore prior to the historic D-Day landing there in June of 1944.

Military orders also called for the creation and training of sailors for permanent naval demolition units. Led by Lieutenant Commander Draper L. Kauffman, who had received the Navy Cross for disarming an intact enemy bomb at Pearl Harbor, allowing the U.S. military to study it, the training program set up at Fort Pierce, Florida, turned out Naval Combat Demolition Units (NCDU) specialists who could map enemy beaches, knock out mines, explode obstacles, and open the way for Allied assaults in the most dire combat.

In the Pacific Theater of Operations, Marines landing on islands and atolls required a different type of assistance than that of the European Theater.

At Tarawa Atoll in the Gilbert Islands U.S. Marines had been met, in November of 1943, with 4,500 well-fortified, well-prepared Japanese soldiers in a bloody scramble to secure that piece of land. The battle left 6,400 Americans, Koreans, and Japanese dead. The Americans gained control of the atoll, but the Japanese had fought down to the man in a show

⋩ **Marine demolition teams storm a Japanese stronghold on Betio Island in the Tarawa Atoll of the Gilbert Islands, November 1943.**

of resistance not seen in earlier amphibious landings. It was clear that in the push to control key islands in the Pacific Marines would need underwater reconnaissance and demolition of any obstructions, particularly the coral reefs that crowded the coastal waters. There was also the possibility of Japanese mines.

The first Underwater Demolition Teams (UDTs), One and Two, were created when 30 officers and 150 enlisted men from the SeaBees, Marines, Army Engineers, and others who had trained at Fort Pierce, were brought to Waimanalo Amphibious Training Base on Oahu,

Hawaii, and later to a Naval Combat Demolition Training and Experimental Base on Maui. For Pacific missions they had to know how to graph the coastline and blast. And they had to be able to swim long distances. Earlier nighttime reconnaissance had been a failure, so the UDTs had to perform their missions in daylight, and that required strong swimmers, not equipment.

Often in nothing more than swim suits, fins, and dive masks, these brave men participated in every significant amphibious landing in the Pacific—Eniwetok, Saipan, Guam, Leyte, Iwo Jima, Okinawa, Brunei Bay, among others.

★★★★★

The story of the United States Navy's frogmen is a story of adventure. Of brave men against the enemy . . . and against the sea. The work they did in the Pacific in World War II, and later in the waters of Korea . . . stamps the underwater demolition team sailors as giants of physical strength, and towers of moral and physical courage. The average frogman is not a giant. What is it then that makes a UDT man? Watch. We'll show you.

Corny as the voice-over narrative from the U.S. Navy's 1957 documentary *The Navy Frogmen* sounds, the message, even today, is not far off. Today's Naval Special Warfare operators, or SEALs—for SEa, Air and Land, which describes the environments in which they operate—the men responsible for some of the most difficult and dangerous missions have their genesis in the early unconventional warfare units of World War II.

Beginning in 1950 and throughout the Korea Wars, UDTs distinguished themselves, performing beach reconnaissance, channel marking, mine sweeping, demolition, and going behind enemy lines to destroy railroads and tunnels in key port and coastal cities such as Inchon, Taechon, Wonsan, and Hungnam. A decade later Underwater Demolition Teams would travel up the Mekong River deep into

⌃ **Members of U.S. Navy SEAL Team One move down the Bassac River in a SEAL Team Assault Boat (STAB) during operations along the river south of Saigon, November 1962.**

Credit: J.D. Randal, JO1, Department of Defense, Department of the Navy, Naval Photographic Center.

Laos to deliver small watercraft during the Vietnam War.

By spring of 1961 President John F. Kennedy, understanding that the conflict in Southeast Asia was unlike any previous conventional war, decided there was a need for a Naval special force like that of the Army's Green Berets or Special Forces, whose roots lay in World War II's Office of Strategic Services (OSS), and who were highly trained in guerilla warfare. In a speech to Congress on May 25, Kennedy outlined his plans to put man on the moon, and to "expand rapidly and substantially, in cooperation with our Allies, the orientation of existing forces for the conduct of nonnuclear war, paramilitary operations and sub-limited or unconventional wars."

The newly reoriented forces would operate in sea, air, and on land. Many would come from the UDTs that had already seen combat in Korea. Two SEAL teams, based on opposite coasts, Team One in Coronado, California, and Team Two in Little Creek, Virginia, would add to their roster of combat skills, high-altitude parachuting, hand-to-hand combat, the operation of newly designed assault rifles, safecracking, demolition, platoon tactics, and foreign languages. Deployed to Vietnam in 1962 to train Vietnamese UDTs—the *Lien Doc Nguoi Nhia*—roughly translated, the "soldiers who fight under the sea," by early 1963 SEALs were engaged in covert operations directed by the Central Intelligence Agency (CIA) against North Vietnamese Army personnel and Vietcong sympathizers. No longer a conflict where soldiers fired artillery at a proscribed target, American soldiers fought within hand's reach of the enemy, guerilla-style. Just as the enemy hid in the dense and tangled landscape of Southeast Asia, so too did the SEALs, painting themselves with camouflage and melting into the jungle. The Vietcong would call the SEALs "The Men with Green Faces."

≈ **Men with Green Faces: Navy SEALs X-Ray Platoon. Taken on a dock near Ben Tre in Southeast Vietnam, 1970.**

1

WHO ARE THESE AMERICANS?

In October 2011, two relief workers, American Jessica Buchanan and Poul Hagen Thisted were kidnapped at gunpoint by Somali pirates, betrayed by their security adviser who had arranged for them to be abducted. For three months they were held hostage while the pirates demanded ransom. Fed only enough to be kept alive and forced to sleep on mats in the open desert, Buchanan fell ill from a thyroid condition that caused a serious kidney ailment. Stuck in a compound, the hostages had no way of knowing whether anyone even knew they had been kidnapped, let alone whether the ransom would be paid.

★★★★★

The pirates around her were dead-to-the-world asleep when Buchanan rose to go to the bathroom in the bush. She had no sooner laid down again when she heard gunfire followed by voices—American voices—calling her name, Jessica, and yanking her out of sleep. The voices don't jibe with what she knows: She is in Somalia in a compound. She has been a hostage of Somalian pirates for three months. She is very ill.

She pulls her blanket more snugly over her, but the voices persist, telling her that they are with the American military, that she is safe, and that they are going to take her home.

With no time to spare, one of the military men picks her up and begins to run. It is pitch dark. Eventually he stops and puts her down. They give her food, water, and medicine, and form a ring around her to keep her safe until the helicopters come. At one point, one of the men runs back to get a ring that belonged to her mother that she had hidden—in the hubbub she had left it behind.

Jessica flew off in one helicopter; the members of SEAL Team Six left in another. She never saw their faces and never knew their names, but to these men who did not know her, this young woman was the most precious thing at that moment, her life more important than their own.

★★★★★

When members of SEAL Team Six shot and killed Osama bin Laden inside a private residential compound in Abbottabad, Pakistan,

≈ **U.S. Navy SEALs with Special Operations Task Force-South shield themselves from dust and rocks as an MH-47 Chinook takes off after a clearing operation in the Panjwai district of Kandahar province, Afghanistan.**
Credit: U.S. Army photo by Sgt. Daniel P. Shook/Released.

on May 2, 2011, the news electrified the world and millions watched as President Obama made the announcement that the man responsible for the World Trade Center attack on 9/11 was dead. It is more than likely that fewer people followed the story of Jessica Buchanan.

In January 2012, some dozen men from Seal Team Six parachuted from an Air Force Special Operations aircraft to a location two miles from where Buchanan, 32, and Thisted, 60, who had been working for a Danish relief organization helping to remove mines in war-torn areas in

≈ **During an interdiction operations exercise in the Arabian Sea, U.S. Navy SEALs fast-rope to the bridge-wing aboard USS *Shreveport* from an HH-60H Seahawk helicopter assigned to the Dragon Slayers of Helicopter Anti-Submarine Squadron One. The exercise simulates boarding a ship carrying terrorist suspects.**

Somalia, were being held. The commandos made their way to the compound in the dark, and surprised the captors, who had been asleep. For Buchanan, it was a miracle that so much military manpower would be exerted to save a lowly aid worker. But to the SEALs, the mission to save her was as important as any, and only one of the many that come under the designation "Special Operations"—what we often hear called Special Ops or SPECOPS.

The U.S. military describes special operations as "actions conducted by specially organized, highly trained, and equipped military and

paramilitary forces to achieve military, political, economic, or psychological objectives by non-conventional military means in hostile, denied, or politically sensitive areas." Men like the Navy SEALS who are part of the Special Operations Forces (SOF) have the specialized skills and training to perform missions that are well beyond those of the regular military.

The risks are higher for special operatives—they often work clandestinely, covertly, and independently from friendly support: The secrecy of their missions makes it so that they are often on their own, with no one but each other and their special training to rely on if

U.S. Air Force pararescuemen and Navy SEALs leap from the ramp of an Air Force C-17 transport aircraft during free-fall parachute training over Marine Corps Base Hawaii.

Credit: U.S. Marine Corps photo by Lance Cpl. Reece E. Lodder.

stuck behind enemy lines, or in a sticky situation in a friendly or allied nation. Special operatives must be highly skilled at understanding and interpreting detailed, high-level intelligence—one minor misstep in communication could mean disaster. There is often no ground or air support to cover them as they carry out their missions.

In fact, special operatives are often at the vanguard, performing reconnaissance and other tasks for marine landings

U.S. Navy SEALs ride in a Navy Special Warfare humvee to provide security for a simulated prisoner escort during a capabilities exercise at Joint Expeditionary Base Little Creek-Fort Story.

Credit: U.S. Navy photo by Petty Officer 3rd Class James Ginther.

and the like. They are often the first to encounter obstacles, snipers, enemy fire, booby traps, and other dangerous impediments, sometimes from the unlikeliest source.

During Operation Restore Hope in Somalia in 1992, Seal Team One swam to shore at Mogadishu to measure beach compositions, water depth, and shore gradient. What they didn't know was that the water was full of waste. Although the team completed their mission, some of the men became ill. Special Ops teams must be flexible, must be able to improvise, must be able to deal with the unexpected, or change plans at a moment's notice. Special Op forces can be subject to the whim of indigenous "assets"— natives or citizens of a foreign country working with the special operatives—so they must know

≋ A SEAL combat diver during daytime training operations. Such operations would normally be conducted during periods of darkness.

Credit: U.S. Navy photo by Senior Chief Mass Communication Specialist Andrew McKaskle.

≋ Navy SEAL Team members secure a U.S. Embassy as part of a noncombatant evacuation exercise during Desert Rescue XI at Naval Air Station Fallon, Nevada. The exercise simulates the rescue of downed aircrew behind enemy lines enabling other aircrews to perform Combat Search and Rescue related missions as well as experiment with new techniques in realistic scenarios. Desert Rescue XI is a joint service training exercise hosted by the Naval Strike and Warfare Center.

cultural customs and beliefs and foreign languages. These are just some of the skills and training required to become a special operative—a Navy SEAL.

★★★★★

"These recommendations will adapt and reshape our defense enterprise so that we can continue protecting this nation's security in an era of unprecedented uncertainty and change. . . . We are repositioning to focus on the strategic challenges and opportunities that will define our future: new technologies; new centers of power; and a world that is growing more volatile, more unpredictable,

⌃ Large munitions cache discovered in the Zhawar Kili area of Eastern Afghanistan, January 2002. Over 70 caves were inspected by SEAL and Navy EOD members before being destroyed by set charges or air strikes.

and in some instances more threatening to the United States. . . . This required a series of difficult choices. We chose further reductions in troop strength and force structure in every military service—active and reserve—in order to sustain our readiness and technological superiority and to protect critical capabilities like special operations forces [including Navy SEALs] and cyber resources. . . . Our recommendations seek to protect capabilities uniquely suited to the most likely missions of the future, most notably special operations forces used for counterterrorism and crisis response. Accordingly, our special operations forces will grow to 69,700 personnel from roughly 66,000 today."

—Secretary Hagel on the fiscal year 2015 Department of Defense budget preview, Pentagon Briefing Room, February 24, 2014

★★★★★

What was written about SEAL missions in the preceding pages are just some examples of the better-known operations. For every one mission that you might hear about in the media, there are perhaps scores—hundreds—that have been conducted in complete secrecy. Most of them come under the categories of counterterrorism, direct action, unconventional warfare, foreign internal defense, or special reconnaissance. SEALs are now operating in 89 different countries.

According to Joint [Chiefs of Staff] Publication 3-05 Special Operations dated April 18,

2011, *Unconventional Warfare* (UW) is defined as "activities conducted to enable a resistance movement or insurgency to coerce, disrupt, or overthrow a government or occupying power by operating through or with an underground, auxiliary, and guerrilla force in a denied area."

The operations are often long-term, and can include sabotage, guerilla warfare, intelligence activities, subversion, and unconventional assisted recovery. With its roots in guerilla activity behind enemy lines during World War II, UW has been employed along with conventional warfare in such operations as Enduring Freedom in Afghanistan in 2001 and Iraqi Freedom in 2003.

Foreign Internal Defense (FID) is "participation by civilian and military agencies of a government in any of the action programs taken by another government or other designated organization to free and protect its society from subversion, lawlessness, insurgency, terrorism, and other threats to its security." FID is overt and direct assistance to the host nation's regular forces to secure and free that nation from said threats. In Iraq and Afghanistan, SEALs trained citizens to perform security and military operations in this capacity.

Counterterrorism (CT) is "actions taken directly against terrorist networks and indirectly to influence and render global and regional

⋏ **A member of the Armed Forces of the Philippines Naval Special Operations Group participates in a battlefield exercise during a combat medic subject matter expert exchange at Naval Base Cavite, Philippines.**

environments inhospitable to terrorist networks." In environments where conventional forces cannot operate because of political situations or threats, SOFs may be employed.

Direct Action (DA) is "strikes and other small-scale offensive actions conducted as a special operation in hostile, denied, or diplomatically sensitive environments, and which employ specialized military capabilities to seize, destroy, capture, exploit, recover, or damage designated targets." Special forces may conduct raids, ambushes and sabotage missions and may have to engage in close quarters combat, ship

⌃ **During a search-and-destroy mission, U.S. Navy SEALs discover a large cache of munitions in one of the many caves in the Zhawar Kili area. Used by al Qaeda and Taliban forces, the caves and above ground complexes were subsequently destroyed through air strikes called in by the SEALs.**

boarding, and seizure. The operation to save the life of Capt. Richard Phillips of the *Maersk Alabama* from Somali pirates who had taken him hostage is a clear example of a direct action. Three of the four pirates holding Phillips captive were shot and killed by Navy SEAL snipers.

Special Reconnaissance (SR) is "reconnaissance and surveillance actions conducted as a special operation in hostile, denied, or politically sensitive environments to collect or verify information of strategic or operational significance,

« **A U.S. Navy SEAL climbs a ladder during a ship assault training scenario. Navy SEALs are a part of a continuous training cycle to improve and further specialize their skills.**
Credit: U.S. Navy photo by Mass Communication Specialist 2nd Class William S. Parker.

employing military capabilities not normally found in conventional forces."

Are You Tough Enough?

You have to be in excellent physical condition to become a Basic Underwater Demolition/ SEAL (BUD/S) candidate so that you can make it through the difficult and uncompromising selection and training process. But there's more. Being physically tough is essential, but without mental resilience and the right mindset—you will not make it.

U.S. Navy SEALs talk to local Afghanis while conducting a sensitive site exploitation mission in the Jaji Mountains. »
Credit: U.S. Navy photo by Photographer's Mate 1st Class Tim Turner.

BUD/S, which will be discussed in detail in chapter 3, is designed to push the trainees well beyond their normal "self perceived" physical limits and to identify those truly qualified to serve in the SEAL teams. To ensure that only those individuals who have the mental toughness are selected, BUD/S is also a test of the mind. As a matter of fact, the SEAL instructors do break down all of the trainees physically, but the ones that they cannot break down mentally most often succeed and graduate the arduous training course. BUD/S is considered as the toughest military training in the world and produces the world's finest warriors.

Trainees are bombarded with a constant barrage of stressors, one more difficult than the next, forcing most of them to drop out. Most of the trainees are not cut out to be SEALs, and go into other career fields in the Navy. BUD/S training is much more intense than most trainees ever imagine.

All members of a SEAL team rely upon one another with complete trust and confidence in their actions and abilities. Every SEAL possesses a loyalty to his team rarely seen in other communities. This team loyalty is required in order for a team to carry out the most dangerous and difficult of missions. There is no weak link in a SEAL team—their lives, and the lives of others depend on this.

≈ BUD/S trainees lift a large log nicknamed Old Misery during log physical training at the Naval Special Warfare Center at Naval Amphibious Base in Coronado, California.

This is why mental resilience is so vastly important and is tested so rigorously in BUD/S training. But what is it exactly?

"Mental toughness is many things and rather difficult to explain. Its qualities are sacrifice and self-denial. Also, most importantly, it is combined with a perfectly disciplined will that refuses to give in. It's a state of mind—you could call it 'character in action.'"

—Vince Lombardi

Think of Capt. Chesley Sullenberger who successfully ditched US Airways Flight 1549 in the Hudson River off New York City in 2009

⌃ **A member of a U.S. Navy SEAL team provides cover for his teammates while advancing on a suspected location of al-Qaeda and Taliban forces in Eastern Afghanistan.**

after Canada geese flew into the engines, disabling the plane. All 155 passengers and crew walked off that plane. A former air force fighter pilot, Sullenberger had 40 years and 20,000 hours of flying experience. He was an instructor and airline accident investigator, and had a reputation as an expert in airline safety.

In January of 2009 he brought that expertise to bear by performing one of the most difficult maneuvers a pilot can make—it is not called "ditching" for nothing. He kept his cool in an extremely stressful situation, and in the process helped his crew perform flawlessly. Coolness under pressure is one of the characteristics of mental toughness.

In October 2012, Malala Yousafzai, a 15-year-old Pakistani schoolgirl who had spoken out against the Taliban and become a global activist for women's rights and education, was shot in the head and neck in an attempt by the Taliban to assassinate her. This did not stop her: On the contrary, this has made her more determined to fight. But what was most striking about Malala was her smile from her hospital bed.

In the wake of her horrific attack (two other girls were also wounded), Malala, who was nominated for a Nobel Peace Prize, remains positive and focused. She is attending school in Birmingham, England, and spoke at the United Nations in July 2013, even though the Taliban reiterated their intention to kill her:

Dear brothers and sisters, do remember one thing. Malala day is not my day. Today is the day of every woman, every boy and every girl who have raised their voice for their rights. There are hundreds of Human rights activists and social workers who are not only speaking for human rights, but who are struggling to achieve their goals of education, peace and equality. Thousands of people have been killed by the terrorists and millions have been injured. I am just one of them. So here I stand . . . one girl among many.

The ability to remain positive and focused is also part of what makes a person mentally tough. They are not distracted no matter what the setbacks, or difficult circumstances. They keep their eyes on the prize.

Captain Sullenberger and Malala Yousafzai and their stories are widely known. They are heroes to many. But some of your most mentally tough people are ordinary folk with extraordinary powers to persevere through crushing illness or crisis—a cancer patient who goes to all of the difficult, often debilitating chemotherapy, even though the odds are against him. A person who meets a terminal illness with courage. A pediatric surgeon who must perform the most delicate microsurgery on a newborn to save its life. An earthquake victim who survives for more than two weeks buried under rubble. Some of what these people have defines the mental toughness necessary to be a SEAL.

Another thing they have is an ability to control their fear. As a SEAL, you will find yourself in dangerous circumstances where you must perform a variety of tasks, often with high-tech equipment and weapons, in difficult climates or weather, all with the objective of completing your mission. And often while being shot at. You will have to learn, through training, how to anticipate the fear you will feel—and unless you are a cyborg you *will* feel fear. The BUD/S program will condition candidates to anticipate, control, and neutralize fear.

★★★★★

Now, the Navy said to me, they said that if you join, we're going to pay you $1,332.60 per month. And they said, if you join, we guarantee that you will have zero minutes per day of privacy in your first few months.

⌃ **Ultimate Mental Toughness: Retired BMC (SEAL) Kenneth J. Stethem receives a bronzed master chief cover from ISC Dan Mayfield and CTTC Jeff Kuhlman, both assigned to USS *Stethem* (DDG 63), during a ceremony to celebrate his brother and ship namesake, Robert D. Stethem's, honorary frocking to the rank of master chief petty officer in 2010 in Yokosuka, Japan. Steelworker 2nd Class Robert Dean Stethem was killed 25 years earlier during the hijacking of TWA Flight 847. He was singled out by Lebanese hijackers because of his military status and was beaten and killed after the terrorists' demands were not met. Throughout his ordeal, Petty Officer Stethem did not yield; instead he acted with fortitude and courage and helped his fellow passengers to endure by his example.**

Credit: U.S. Navy photo by Mass Communication Specialist 1st Class Geronimo Aquino.

And they said, if you join, from the moment that you sign your name on the dotted line, you're going to owe us eight years of service. In return, we'll give you one and only one chance at Basic Underwater Demolition Seal (BUD/S) School. If you make it, then you're going to be on your way to being a SEAL officer and leader. But if you fail, as over 80 percent of the candidates do, then you're still going to owe us eight years and we'll tell you where and how you're gonna serve.

Now, in some way, it wasn't such a compelling offer. But when I thought about that question in my own life, I realized that a university career at Oxford could give me a lot of freedom. The consulting firm could give me a lot of money, the Navy was going to give me very little [money], but would make me [into something] more. And they were going to give me the opportunity to be of service to others and to lead and to test myself. And so I think that whenever you or whenever we are facing these decisions in our lives, if we ask ourselves the question, which decision, which path on our journey is going to make us more, I think we'll always be happy with the choices that we make.

—Eric Greitens, Navy SEAL and
bestselling author

★★★★★

Navy SEAL Eric Greitens passed up the academic life at Oxford University and a lucrative consulting job to become a Navy SEAL, with no guarantee that he would make it through

BUD/S. But clearly, the rewards for him have been big.

The question you must ask yourself is why you want to be a Navy SEAL, and not join another branch of the military, or even join the military at all? This is something to think about—this endeavor should not be begun on a whim or taken lightly, or because your buddies are doing it, but because you know, in your heart, mind, and soul that you have what it takes, and more importantly, that this is right for you and you are right for it!

In the next chapter "Know What You're Getting Into," we'll outline the various commitments involved in becoming a SEAL—and there are many.

2

KNOW WHAT YOU'RE GETTING INTO

Now that you've read a bit of Navy SEAL history and determined that you might have what it takes to become one, it's time for the next step—getting down to specifics—the qualifications needed just to be given a chance to make it to BUD/S and for those few who make it to the teams. But there is something that you need to know.

It is this:

Paragraph 10 of the enlistment contract below states a) FOR ALL ENLISTEES: If this is my initial enlistment, I must serve a total of eight (8) years. Any part of that service not served on active duty must be served in a Reserve Component unless I am sooner discharged.

Eight years. Minimum. So, if you do four years of active duty and "get out" you are still in the military, only now in the inactive reserves, officially the Individual Ready Reserves, which means that the military can call you up at any time in the remaining four years.

What if you graduate from a service academy or ROTC program? Five years is the standard commitment, but graduates who accept pilot training are committed to active duty for nine years. Other active-duty commissioning programs usually require a minimum of three years.

★★★★★

When you sign up with a branch of military service you sign an enlistment contract that outlines your initial commitment, bonuses, job training guarantees and other incentives. While this may not be the most exciting part of this handbook, it may be the most important. Enthusiasm will get you to a recruiter, but when you sign that contract you must do so with your eyes wide open and with full knowledge of your obligations as well as your compensation. As for the compensation, SEALs do not become SEALs for the income and benefits.

Eight years may seem like it flies by if you have made the right choice for yourself by joining the military, or it can drag on if, in the end, you didn't, and have to stick it out. The time commitment to the military is life changing. But during your time in the military you will also be subject to another life altering event, the Permanent Change of Station (PCS).

The PCS is "the official relocation of an active duty military service member—along with any family members living with him or her—to a different duty location, such as a military base." This commitment lasts until another PCS order is issued, upon completion of active duty service, or by some other preemptive event.

In other words, you will move, and you may move a lot. And sometimes far away from extended family and friends. Your kids may have to attend several different schools during your active duty service. If there is no space on the military base, you will likely have to find a home to rent or buy in a nearby town. You and your family may live in a place that is not geographically desirable. If you serve more than two years, a PCS move may require you to extend your enlistment because moving a service member costs the government money. Most SEALs are stationed in Coronado, California, or Virginia Beach, Virginia; however, they are normally training or deployed away from home up to three hundred days a year.

★★★★★

All of the branches of service use the same enlistment contract (see sample on the pages that follow). This contract is used for military enlistments and reenlistments. You will sign various paperwork when joining the military, but this is the most important. The basic rule is this: If it's not in your active duty contract then it doesn't matter what the recruitment officer told you. The active duty contract is the *second* contract you'll sign. The first contract you'll sign is called a Delayed Enlistment Program (DEP) contract.

This is an enlistment into the inactive reserves, with an agreement to report for active duty (to ship out to boot camp) at a specific time in the future. A recruit can remain in the DEP for up to 365 days.

You will be "sworn" into the DEP. This reservation is nonbinding, a recruit can walk away from it at any time, but it is a way to solidify the recruit's commitment in by way of ceremony.

The moment that you sign your final contract, the one you sign right before you ship out to Initial Entry Training (IET), which includes both Basic and Advanced Individual Training (AIT), and you take the Oath of

Enlistment, your active duty begins. Entitlements such as medical care, base pay, and the G.I. Bill are allowed by law and are not included in the contract because these benefits are available to everyone who enlists in the military. Any

special incentive, and to which your rank is entitled—enlistment bonus, college loan repayment program, advanced rank—must be in that final contract, or you will not receive them.

It is possible to get out of your enlistment contract but it is more difficult to "walk away." Still, many enlistees "wash out" of boot camp and return home with some sort of General Discharge (which may or may not be under honorable conditions). Otherwise, the degree of difficulty surrounding getting out of your contract depends on the needs of the nation and the availability of talent in your chosen field. Not completing your commitment will not look good to future civilian employers. Simply put, you should think long and hard before you sign an enlistment contract, and if you do, you should plan on fulfilling that commitment.

Family

Another very important thing to think about when deciding on whether a future as a Navy SEAL is for you is family. If you are an enlisted man you will have to plan on a six-year commitment at the minimum. When you combine all of the time that will be taken to get through boot camp, PRE-BUD/S, and BUD/S it is more than sixty-five weeks away from your home and loved ones. And that is just the start.

There is normally leave or vacation time after you earn your Trident, followed by deployment with your first command and/or more advanced training, and then up to nine months on assignment. All military personnel are allotted 30 days of leave per year.

SEAL officers have a minimum ten-year commitment, with one to two summers of Officer Candidate School during or after college, for some the Naval Academy, for others a traditional university. And of course there will be time off from school.

The SEAL officer attends BUD/S with the enlisted men. After graduation, officers are obligated to serve for seven years.

You will have to accept that you will have limited contact with your family during recruit and SEAL candidate training right up to graduation.

For some recruits and SEAL candidates, it will be the first time they have spent such long periods of time away from families, and the separation is equally difficult for parents and other loved ones. Family members can find support by connecting with other SEAL candidate families.

Thinking full-time active duty is not for you, but still want to serve? Consider the

---

For Mom and Dad, Teachers, Guidance Counselors, Coaches, Friends, and Family

Your son, student, or loved one is making a decision that, if all works out, will be a serious commitment of body and soul, and it comes in stages. At each stage, from enlistment or commissioning, to BUD/S to active duty as a Navy SEAL, things will become more challenging. The best thing that you can do is to give him as much support as possible.

Educating yourself about the SEALs will not only lay the groundwork for conversation, it will help you to better understand what your son, loved one, or student is going through.

And remember, this is not just life-changing for him. You will have your own challenges, particularly that of your son, loved one, or student being away from home or incommunicado for long stretches at a time. Understanding what will be required of him will help quell your own concerns and fears.

Research online at official SEAL websites including sealswcc.com, navy.com, and navy.mil. Take notes and create talking points for later conversations with your son, loved one, or student.

Connect with other parents of recruits and SEAL candidates; you will soon find as he progresses that you have not only made new friends but will also become a part of a new, exclusive SEAL family.

Most of you will have a good idea about a young man's desire for a military career well before he approaches you with his wish to become a SEAL. But there will be those to whom this knowledge will come as quite a shock. Ask him why he wants to become a SEAL. Help him understand what a commitment to the U.S. military means. Your own experience will come to bear in pointing out aspects of the commitment that he might not have thought about.

Talk to your local Navy recruiter or SEAL scout together. You can call 888-USN-SEAL (888-876-7325) for more information.

Reserves or National Guard. Your obligation is generally one weekend a month, plus two weeks of active duty a year. As an enlisted man, after the initial seven weeks of boot camp at Recruit Training Command at Great Lakes, Illinois, you can train near home until called to active duty, and pursue a full-time civilian education or obtain special military training while serving.

ENLISTMENT/REENLISTMENT DOCUMENT
ARMED FORCES OF THE UNITED STATES
PRIVACY ACT STATEMENT

AUTHORITY: 5 U.S.C. 3331; 10 U.S.C. 113, 136, 502, 504, 505, 506, 507, 508, 509, 510, 513, 515, 516, 518, 519, 972, 978, 2107, 2107a, 3253, 3258, 3262, 5540, 8252, 8253, 8257, 8258, 12102, 12103, 12104, 12105, 12106, 12107, 12108, 12301, 12302, 12304, 12305, 12405; 14 USC 351, 632; 32 U.S.C. 301, 302, 303, 304; and Executive Order 9397, November 1943 (SSN).

PRINCIPAL PURPOSE(S): To record enlistment or reenlistment into the U.S. Armed Forces. This information becomes a part of the subject's military personnel records which are used to document promotion, reassignment, training, medical support, and other personnel management actions. The purpose of soliciting the SSN is for positive identification.

ROUTINE USE(S): This form becomes a part of the Service's Enlisted Master File and Field Personnel File. All uses of the form are internal to the relevant Service.

DISCLOSURE: Voluntary; however, failure to furnish personal identification information may negate the enlistment/reenlistment application.

A. ENLISTEE/REENLISTEE IDENTIFICATION DATA

1. NAME *(Last, First, Middle)*		2. SOCIAL SECURITY NUMBER			
3. HOME OF RECORD *(Street, City, County, State, Country, ZIP Code)*		4. PLACE OF ENLISTMENT/REENLISTMENT *(Mil. Installation, City, State)*			
5. DATE OF ENLISTMENT/ REENLISTMENT *(YYYYMMDD)*	6. DATE OF BIRTH *(YYYYMMDD)*	7. PREV MIL SVC UPON ENL/REENLIST	YEARS	MONTHS	DAYS
		a. TOTAL ACTIVE MILITARY SERVICE			
		b. TOTAL INACTIVE MILITARY SERVICE			

B. AGREEMENTS

8. I am enlisting/reenlisting in the United States *(list branch of service)* _____
this date for _____ years and _____ weeks beginning in pay grade _____ of which
_____ years and _____ weeks is considered an Active Duty Obligation, and _____ years and
_____ weeks will be served in the Reserve Component of the Service in which I have enlisted. If this is an initial enlistment, I must serve a total of eight (8) years, unless I am sooner discharged or otherwise extended by the appropriate authority. This eight year service requirement is called the Military Service Obligation. The additional details of my enlistment/ reenlistment are in Section C and Annex(es) *(list name of Annex(es) and describe)*
_____
_____

a. FOR ENLISTMENT IN A DELAYED ENTRY/ENLISTMENT PROGRAM (DEP):
I understand that I am joining the DEP. I understand that by joining the DEP I am enlisting in the Ready Reserve component of the United States *(list branch of service)* _____ for a period not to exceed 365 days, unless this period of time is otherwise extended by the Secretary concerned. While in the DEP, I understand that I am in a nonpay status and that I am not entitled to any benefits or privileges as a member of the Ready Reserve, to include, but not limited to medical care, liability insurance, death benefits, education benefits, or disability retired pay if I incur a physical disability. I understand that the period of time while I am in the DEP is NOT creditable for pay purposes upon entry into a pay status. However, I also understand that the period of time while I am in the DEP is counted toward fulfillment of my military service obligation described in paragraph 10, below. While in the DEP, I understand that I must maintain my current qualifications and keep my recruiter informed of any changes in my physical or dependency status, qualifications, and mailing address. I understand that I WILL be ordered to active duty unless I report to the place shown in item 4 above by *(list date (YYYYMMDD))* _____ for enlistment in the Regular component of the United States *(list branch of service)* _____
for not less than _____ years and _____ weeks.
b. REMARKS: *(If none, so state.)*

c. The agreements in this section and attached annex(es) are all the promises made to me by the Government. **ANYTHING ELSE ANYONE HAS PROMISED ME IS NOT VALID AND WILL NOT BE HONORED.**
(Initials of Enlistee/Reenlistee) _____
(Continued on Page 2)

DD FORM 4/1, OCT 2007   PREVIOUS EDITION IS OBSOLETE.   Adobe Professional 8.0

C. PARTIAL STATEMENT OF EXISTING UNITED STATES LAWS

9. FOR ALL ENLISTEES OR REENLISTEES:
I understand that many laws, regulations, and military customs will govern my conduct and require me to do things under this agreement that a civilian does not have to do. I also understand that various laws, some of which are listed in this agreement, directly affect this enlistment/reenlistment agreement. Some examples of how existing laws may affect this agreement are explained in paragraphs 10 and 11. I understand that I cannot change these laws but that Congress may change these laws, or pass new laws, at any time that may affect this agreement, and that I will be subject to those laws and any changes they make to this agreement. I further understand that:

a. My enlistment/reenlistment agreement is more than an employment agreement. It effects a change in status from civilian to military member of the Armed Forces. As a member of the Armed Forces of the United States, I will be:

(1) Required to obey all lawful orders and perform all assigned duties.

(2) Subject to separation during or at the end of my enlistment. If my behavior fails to meet acceptable military standards, I may be discharged and given a certificate for less than honorable service, which may hurt my future job opportunities and my claim for veteran's benefits.

(3) Subject to the military justice system, which means, among other things, that I may be tried by military courts-martial.

(4) Required upon order to serve in combat or other hazardous situations.

(5) Entitled to receive pay, allowances, and other benefits as provided by law and regulation.

b. Laws and regulations that govern military personnel may change without notice to me. Such changes may affect my status, pay, allowances, benefits, and responsibilities as a member of the Armed Forces **REGARDLESS** of the provisions of this enlistment/reenlistment document.

10. MILITARY SERVICE OBLIGATION, SERVICE ON ACTIVE DUTY AND STOP-LOSS FOR ALL MEMBERS OF THE ACTIVE AND RESERVE COMPONENTS, INCLUDING THE NATIONAL GUARD.

a. FOR ALL ENLISTEES: If this is my initial enlistment, I must serve a total of eight (8) years, unless I am sooner discharged or otherwise extended by the appropriate authority. This eight year service requirement is called the Military Service Obligation. Any part of this service not served on active duty must be served in the Reserve Component of the service in which I have enlisted. If this is a reenlistment, I must serve the number of years specified in this agreement, unless I am sooner discharged or otherwise extended by the appropriate authority. Some laws that affect when I may be ordered to serve on active duty, the length of my service on active duty, and the length of my service in the Reserve Component, even beyond the eight years of my Military Service Obligation, are discussed in the following paragraphs.

b. I understand that I can be ordered to active duty at any time while I am a member of the DEP. In a time of war, my enlistment may be extended without my consent for the duration of the war and for six months after its end (10 U.S.C. 506, 12103(c)).

c. As a member of a Reserve Component of an Armed Force, in time of war or of national emergency declared by the Congress, I may, without my consent, be ordered to active duty, for the entire period of the war or emergency and for six (6) months after its end (10 U.S.C. 12301(a)). My enlistment may be extended during this period without my consent (10 U.S.C. 12103(c)).

d. As a member of the Ready Reserve (to include Delayed Entry Program), in time of national emergency declared by the President, I may, without my consent, be ordered to serve on active duty, and my military service may be extended without my consent, for not more than 24 consecutive months (10 U.S.C. 12302). My enlistment may be extended during this period without my consent (see paragraph 10g).

e. As a member of the Ready Reserve, I may, at any time and without my consent, be ordered to active duty to complete a total of 24 months of active duty, and my enlistment may be extended so I can complete the total of 24 months of active duty, if:

(1) I am not assigned to, or participating unsatisfactorily in, a unit of the Ready Reserve; and

(2) I have not met my Reserve obligation; and

(3) I have not served on active duty for a total of 24 months (10 U.S.C. 12303).

f. As a member of the Selected Reserve or as a member of the Individual Ready Reserve mobilization category, when the President determines that it is necessary to augment the active forces for any operational mission or for certain emergencies, I may, without my consent, be ordered to active duty for not more than 365 days (10 U.S.C. 12304). My enlistment may be extended during this period without my consent (see paragraph 10g).

g. During any period members of a Reserve component are serving on active duty pursuant to an order to active duty under authority of 10 U.S.C. 12301, 12302, or 12304, the President may suspend any provision of law relating to my promotion, retirement, or separation from the Armed Forces if he or his designee determines I am essential to the national security of the United States. Such an action may result in an extension, without my consent, of the length of service specified in this agreement. Such an extension is often called a "stop-loss" extension (10 U.S.C. 12305).

h. I may, without my consent, be ordered to perform additional active duty training for not more than 45 days if I have not fulfilled my military service obligation and fail in any year to perform the required training duty satisfactorily. If the failure occurs during the last year of my required membership in the Ready Reserves, my enlistment may be extended until I perform that additional duty, but not for more than six months (10 U.S.C. 10148).

11. FOR ENLISTEES/REENLISTEES IN THE NAVY, MARINE CORPS, OR COAST GUARD: I understand that if I am serving on a naval vessel in foreign waters, and my enlistment expires, I will be returned to the United States for discharge as soon as possible consistent with my desires. However, if essential to the public interest, I understand that I may be retained on active duty until the vessel returns to the United States. If I am retained under these circumstances, I understand I will be discharged not later than 30 days after my return to the United States; and, that except in time of war, I will be entitled to an increase in basic pay of 25 percent from the date my enlistment expires to the date of my discharge.

12. FOR ALL MALE APPLICANTS: Completion of this form constitutes registration with the Selective Service System in accordance with the Military Selective Service Act. Incident thereto the Department of Defense may transmit my name, permanent address, military address, Social Security Number, and birthdate to the Selective Service System for recording as evidence of the registration.

(Initials of Enlistee/Reenlistee) _____

DD FORM 4/1 (PAGE 2), OCT 2007

NAME OF ENLISTEE/REENLISTEE *(Last, First, Middle)*	SOCIAL SECURITY NO. OF ENLISTEE/REENLISTEE

D. CERTIFICATION AND ACCEPTANCE

13a. My acceptance for enlistment is based on the information I have given in my application for enlistment. If any of that information is false or incorrect, this enlistment may be voided or terminated administratively by the Government or I may be tried by a Federal, civilian, or military court and, if found guilty, may be punished.

I certify that I have carefully read this document, including the partial statement of existing United States laws in Section C and how they may affect this agreement. Any questions I had were explained to my satisfaction. I fully understand that only those agreements in Section B and Section C of this document or recorded on the attached annex(es) will be honored. I also understand that any other promises or guarantees made to me by anyone that are not set forth in Section B or the attached annex(es) are not effective and will not be honored.

b. SIGNATURE OF ENLISTEE/REENLISTEE	c. DATE SIGNED *(YYYYMMDD)*

14. SERVICE REPRESENTATIVE CERTIFICATION

a. On behalf of the United States *(list branch of service)* _____,
I accept this applicant for enlistment. I have witnessed the signature in item 13b to this document. I certify that I have explained that only those agreements in Section B of this form and in the attached Annex(es) will be honored, and any other promises made by any person are not effective and will not be honored.

b. NAME *(Last, First, Middle)*	c. PAY GRADE	d. UNIT/COMMAND NAME
e. SIGNATURE	f. DATE SIGNED *(YYYYMMDD)*	g. UNIT/COMMAND ADDRESS *(City, State, ZIP Code)*

E. CONFIRMATION OF ENLISTMENT OR REENLISTMENT

15. IN THE ARMED FORCES EXCEPT THE NATIONAL GUARD (ARMY OR AIR):

I, _____, do solemnly swear (or affirm) that I will support and defend the Constitution of the United States against all enemies, foreign and domestic; that I will bear true faith and allegiance to the same; and that I will obey the orders of the President of the United States and the orders of the officers appointed over me, according to regulations and the Uniform Code of Military Justice. So help me God.

16. IN THE NATIONAL GUARD (ARMY OR AIR):

I, _____, do solemnly swear (or affirm) that I will support and defend the Constitution of the United States and the State of _____ against all enemies, foreign and domestic; that I will bear true faith and allegiance to the same; and that I will obey the orders of the President of the United States and the Governor of _____ and the orders of the officers appointed over me, according to law and regulations. So help me God.

17. IN THE NATIONAL GUARD (ARMY OR AIR):

I do hereby acknowledge to have voluntarily enlisted/reenlisted this _____ day of _____, _____ in the _____ National Guard and as a Reserve of the United States *(list branch of service)* _____ with membership in the _____ National Guard of the United States for a period of _____ years, _____ months, _____ days, under the conditions prescribed by law, unless sooner discharged by proper authority.

18.a. SIGNATURE OF ENLISTEE/REENLISTEE	b. DATE SIGNED *(YYYYMMDD)*

19. ENLISTMENT/REENLISTMENT OFFICER CERTIFICATION

a. The above oath was administered, subscribed, and duly sworn to (or affirmed) before me this date.

b. NAME *(Last, First, Middle)*	c. PAY GRADE	d. UNIT/COMMAND NAME
e. SIGNATURE	f. DATE SIGNED *(YYYYMMDD)*	g. UNIT/COMMAND ADDRESS *(City, State, ZIP Code)*

(Initials of Enlistee/Reenlistee) _____

DD FORM 4/2, OCT 2007   PREVIOUS EDITION IS OBSOLETE.

NAME OF ENLISTEE/REENLISTEE *(Last, First, Middle)*	SOCIAL SECURITY NO. OF ENLISTEE/REENLISTEE

F. DISCHARGE FROM/DELAYED ENTRY/ENLISTMENT PROGRAM

20a. I request to be discharged from the Delayed Entry/Enlistment Program (DEP) and enlisted in the Regular Component of the United States *(list branch of service)* _____ for a period of _____ years and _____ weeks. No changes have been made to my enlistment options OR if changes were made they are recorded on Annex(es) _____

_____

which replace(s) Annex(es) _____

_____.

b. SIGNATURE OF DELAYED ENTRY/ENLISTMENT PROGRAM ENLISTEE	c. DATE SIGNED *(YYYYMMDD)*

G. APPROVAL AND ACCEPTANCE BY SERVICE REPRESENTATIVE

21. SERVICE REPRESENTATIVE CERTIFICATION

a. This enlistee is discharged from the Reserve Component shown in item 8 and is accepted for enlistment in the Regular Component of the United States *(list branch of service)* _____ in pay grade _____ .

b. NAME *(Last, First, Middle)*	c. PAY GRADE	d. UNIT/COMMAND NAME
e. SIGNATURE	f. DATE SIGNED *(YYYYMMDD)*	g. UNIT/COMMAND ADDRESS *(City, State, ZIP Code)*

H. CONFIRMATION OF ENLISTMENT OR REENLISTMENT

22a. IN A REGULAR COMPONENT OF THE ARMED FORCES:

I, _____ , do solemnly swear (or affirm) that I will support and defend the Constitution of the United States against all enemies, foreign and domestic; that I will bear true faith and allegiance to the same; and that I will obey the orders of the President of the United States and the orders of the officers appointed over me, according to regulations and the Uniform Code of Military Justice. So help me God.

b. SIGNATURE OF ENLISTEE/REENLISTEE	c. DATE SIGNED *(YYYYMMDD)*

23. ENLISTMENT OFFICER CERTIFICATION

a. The above oath was administered, subscribed, and duly sworn to (or affirmed) before me this date.

b. NAME *(Last, First, Middle)*	c. PAY GRADE	d. UNIT/COMMAND NAME
e. SIGNATURE	f. DATE SIGNED *(YYYYMMDD)*	g. UNIT/COMMAND ADDRESS *(City, State, ZIP Code)*

(Initials of Enlistee/Reenlistee) _____

DD FORM 4/3, OCT 2007   PREVIOUS EDITION IS OBSOLETE.   Reset

BUD/S students carry an inflatable boat along the beach at the Naval Special Warfare Center in Coronado, California.

ON MENTAL TOUGHNESS

A Warrior must remember who/what he represents . . . always.

"Only the dead have seen the end of war."—Plato

Mindset is the most important aspect of your readiness for BUD/S and your chances of making it through BUD/S. Nothing can stop a person with a strong mindset from achieving anything that person sets out to accomplish. And nothing can help the person with a weak mindset. There is no place in the SPECOPS community for anybody with a weak mindset.

When asked "how should I prepare myself for BUD/S" my answer is basically, "do something every day to make yourself stronger, faster, fitter, and/or smarter."

Stronger can come from push-ups, pull-ups, sit-ups, rope climbs, SEAL PT, CrossFit, P90X, and so on. There are literally thousands of ways to become stronger. It is up to you to have the incentive to research and find the right workout program that fits your schedule and training locations. You do need to practice doing pull-ups. If you do not have a pull-up bar available consider making one.

Faster means to become a faster runner concentrating on the four to six mile runs. Soft sand runs are a big part of BUD/S, so if you can, get to a beach, put on your boots, and do timed runs. As for swimming, learn to swim with fins and do two-mile open water swims using the combat side stroke. If you need to, hire a swim coach to help you with your stroke. If you can do a two-mile timed swim, once a week it would be ideal. As in BUD/S every week you want to have a faster time than the week before.

Your goal should be to be able to do more pull-ups, push-ups, and sit-ups than the maximum PST scores. Your goal should be to swim and run faster than the maximum scores.

Get on a serious workout program and monitor and record your progress. This is no game. You are training to enter the world's most selective and elite fraternity. You are hopefully going to be training, fighting, and living with some of the toughest warriors on this planet. You need to do everything you can to be—all that you can possibly be.

Read *The U.S. Navy SEAL Guide to Fitness* and *The U.S. Navy SEAL Guide to Nutrition*. Eat, sleep, and hydrate properly so you get the most out of your workouts. Many BUD/S candidates go to shooting ranges and learn to shoot and assemble and disassemble weapons. They learn about the fundamentals of marksmanship and ballistics. Many attend civilian dive training and parachute training courses. Dive physics is a difficult subject for many people but if you can learn the fundamentals of diving prior to BUD/S you will be ahead of the game. I actually read the *U.S. Navy Diving Manuals* (volumes one and two) prior to attending BUD/S. The more prepared you go into BUD/S and the more you learn about the teams, military history, and related topics, the better prepared you will be.

"The more sweat and tears you put into the training, the less blood you'll shed in time of war."

—BUD/S motto

I would also suggest that you become an EMT, a wilderness EMT, or at least a first responder. You will be required to know basic emergency medicine, and the chances of you needing these skills to save lives will be high.

Learn all you can about the history of the SEAL Teams, Naval Special Warfare, and military history. Pay attention to the world current events—the wars and the troubled areas in this world. Read the books written by Navy SEALs and watch the SEAL and military movies.

The Modern Day Gunslinger is basically a weapon fundamentals and basic tactics book. Most of the lessons in that book I learned while in the SEAL teams. *The Navy SEAL Survival Handbook* is a book of survival lessons, most of which I learned in the SEAL teams and during the Navy Survival, Escape, Resistance, and Evasion course.

There were a few things I did which gave me a great advantage in BUD/S. I trained at least 20 hours a week and competed in endurance races such as marathons and triathlons regularly. When I entered BUD/S I was very fit and thrived on physical challenges. I wanted to be challenged. My attitude was that I "Welcome the Pain" and that "Pain Breeds Strength, Pain is Good, and Extreme Pain is Extremely Good." If things were too difficult I would simply pass out. If I was still conscious then I could deal with more pain. That is the philosophy I have lived by for over 35 years—and it works!

I also used visualization techniques for four years in preparation for BUD/S. I knew BUD/S was going to be extremely difficult, so I spent hours a week visualizing how cold the water would be, how sore I would be after doing thousands of push-ups and sit-ups a day, hundreds of pull-ups a day, and so on. My visualization helped me out immensely. BUD/S was very difficult but my visualization of BUD/S was much worse than it actually was.

I did not have a contingency plan. All I wanted was to become a Navy SEAL. I did not have nor did I want a backup plan. I feel those who get to BUD/S with anything less than absolute dedication—"well if this doesn't work out, I will just be a diver or a medic"—find excuses to quit when things get very difficult in training. I remained 100 percent determined to give BUD/S 100 percent effort 100 percent of the time. There were times that the water was so cold I just wanted to pass out, knowing I would be pulled out of the cold water and revived, but quitting was never an option.

As they say in BUD/S, "it pays to be a winner." In all of your training you want to try to be first at everything. You and your team need to do everything you can to win. When ordered to do soft-sand sprints on the beach, you need to put out 100 percent and have the goal to win every sprint. When it comes to assembling and disassembling weapons blindfolded, you want to finish first; when cleaning the barracks and preparing your uniform and operational gear for inspection, you do not want to have any violations. In other words, you need to do all you can to be as good as you can at everything you do in BUD/S and hopefully in the teams.

When you finish any assigned task, you need to assist your teammates in every way possible. If a teammate is having problems with dive physiology or dive physics, and you have a good understanding of these subjects, it is up to you to spend your off-duty hours helping out your "swim buddy." One of the most important aspects of making it through BUD/S is how you are as a teammate. Your teammates are everything. You need to do all that you can to make your team stronger.

From the moment you walk into the BUD/S compound you are being evaluated very closely and constantly. Your strengths and weaknesses will soon be revealed in the way you present yourself, the appearance of your uniform, your overall attitude, how you treat your teammates, how hard you try at every task, and virtually every other way that you conduct yourself. Your reputation begins the day you raise your right hand and take the oath to serve your country.

The big picture is that the Navy doesn't spend many millions of dollars training young motivated men to become SEALs just so they can wear the coveted SEAL trident or so they get to be part of a community filled with some of the most experienced warriors on the planet. The reason the Navy spends so much time, energy, and resources training folks to become SEALs is because they are investing in a very elite warrior. Once you make it out of BUD/S and into the teams your role is to protect and defend our nation and that often requires going into harm's way and "neutralizing threats"—in nonpolitical terms this means you are trained to kill the enemy. This fact must be strongly emphasized and considered before making the decision to go to BUD/S. Are you the type of person who can commit his life to training and operating while being away from home most of the time and be willing to take the life of an enemy combatant? If you answer yes to these questions, can be in superb physical conditioning before going to BUD/S, and can stay extremely motivated—and lucky—then you do have a chance of becoming a U.S. Navy SEAL.

3

HOW TO BECOME A SEAL CANDIDATE

Becoming a SEAL candidate is the next step on your journey to becoming a member of this elite community. Applicant, aspirant, contender—these are all synonyms for candidate, and you are all of these. You will be applying for a chance to join the military. You are aspiring to be an elite special forces member. You are a contender among many to become one of the special few. But before you can become a SEAL candidate, you will be a candidate *to become a candidate,* and must meet some minimal but crucial requirements before you will even be considered.

- You must be a U.S. citizen.
- You must be 18 to 28 years old (17 with parental permission). Waivers for men ages 29 and 30 are available for

highly qualified candidates. Officer applicants must be at least 19 years of age and commissioned before their 29th birthday. The maximum age limit may be adjusted upward for prior active service on a month-for-month basis up to 24 months. Waivers beyond 24 months will be considered for active or previously enlisted personnel or civilians who possess exceptional qualifications, if they can be commissioned prior to their 35th birthday. Waivers will be considered by the Naval Special Warfare Officer Community Manager or BUPERS 311D.

- You must be a high school graduate, have a GED, or meet the criteria called a High Performance Predictor Profile (HP3), and be proficient in reading, speaking, writing, and understanding the English language.
- You must have a clean record and not be under civil restraint (a restraining order). You must not be a substance abuser or have a pattern of minor convictions or *any* non-minor, misdemeanor, or felony convictions. Waivers can be granted depending on number and severity of convictions, but applicants with lawsuits pending cannot be enlisted without prior approval.

- A SEAL candidate must meet the Armed Services Vocational Aptitude Battery (ASVAB) minimum requirements, which we will go into detail about later in this chapter.
- Uncorrected vision no worse than 20/40 in best eye and 20/70 in worst eye, correctable to 20/25 or better. Color blindness will disqualify you.
- You must be able to obtain a secret security clearance.
- As of this printing, you must be male.

There are more than a few ways to become a SEAL candidate. As a civilian you can request to join the SEALs prior to enlisting through the SEAL Challenge Contract. The SEAL Challenge Contract guarantees you the opportunity to become a SEAL candidate and entitles you to certain bonuses and benefits when you enlist.

If you don't get a SEAL Challenge Contract before enlisting, you can volunteer to take the Physical Screening Test (PST) during the first week of boot camp. If you pass the PST a Naval Special Operation Motivator will interview you. The motivator will then submit a request for you to enter the BUD/S training pipeline.

★★★★★

Tell the Navy recruiter that you want to take the SEAL Challenge *before* you enlist. If you don't have the contract before joining you won't qualify for the same benefits as applicants that have the contract.

⚆ **Recruits—rating candidates for special warfare operator, special warfare boat operator, Navy diver, and explosive ordnance disposal (EOD)—run on an outdoor track at Recruit Training Command in Great Lakes, Illinois.**
Credit: U.S. Navy photo by Mr. Scott A. Thornbloom

More than anything, a candidate must pass the physical screening test (PST) by at least meeting the minimum requirements. Being prepared to just make the minimum standards is not a good plan. You should be prepared and plan on doing extremely well in this simple test. Those who make it to BUD/S will learn that "it pays to be a winner." The PST is really an entrance exam, one you will take more than once. In relative terms to the physical training in BUD/S it is a very simple, not very challenging exam.

Before we review those minimum requirements, you should understand the order of progression to become a SEAL. First you must pass an initial PST.

A qualifying SEAL PST is administered by a Naval Special Warfare coordinator or mentor. Prospective candidates can increase their chances of being selected for BUD/S and succeeding in training by having optimum PST scores. Passing with minimum scores is not a good way to begin your journey into Naval Special Warfare, commonly referred to as NAVSPECWAR.

SEAL MENTORS

A Navy SEAL Mentor is your guide who will help you navigate the process of and understand and fulfill the specific requirements to becoming a Navy SEAL. They will help you train for your PST and will also administer your Delayed Entry Program (DEP) qualifying PST.

You can find a SEAL Mentor in any one of 26 regional areas in the United States. For information about a SEAL Mentor or PST training, call 888-USN-SEAL (888-876-7325). The SEAL + SWCC Scout Team is available Monday through Friday, 6:00 a.m. to 4:00 p.m. Pacific Standard Time, except federal holidays. Or you can email the SEAL + SWCC Scout Team at motivators@navsoc.socom.mil or talk to your local Navy recruiter. SEAL Mentor contact information is given out on a need-to-know basis.

Regional Areas

- Atlanta, Georgia
- Boston, Massachusetts
- Chicago, Illinois
- Columbus, Ohio
- Dallas, Texas
- Denver, Colorado
- Detroit, Michigan
- Houston, Texas
- Jacksonville, Florida
- Los Angeles, California
- Miami, Florida
- Minneapolis, Minnesota
- Nashville, Tennessee
- New Orleans, Louisiana
- New York, New York
- Philadelphia, Pennsylvania
- Phoenix, Arizona
- Pittsburgh, Pennsylvania
- Portland, Oregon
- Raleigh, North Carolina
- Richmond, Virginia
- San Antonio, Texas
- San Diego, California
- San Francisco, California
- Seattle, Washington
- St. Louis, Missouri

To qualify for a SEAL contract, a prospective candidate must meet the minimum physical requirements. Working to achieve optimum fitness standards (and beyond) and to improve your chances at BUD/S is *highly* recommended. You can judge your current level of fitness by taking the SEAL PST Calculator. The SEAL PST Calculator will compare your scores to actual Basic Underwater Demolition/SEAL (BUD/S) entry-level scores and rank you accordingly. Visit www.sealswcc.com.

Prepare as much as you can before you get to boot camp and BUD/S.

PHYSICAL SCREENING TEST WITH MINIMUM, AVERAGE, AND OPTIMUM SCORES

Swim 500-yard breast or side stroke	12:30	10:00	9:30
Push-ups in two minutes	42	79	100
Sit-ups in two minutes	50	79	100
Pull-ups—no time limit	6	11	25
Run 1.5 miles	11:00	10:20	9:30

Points of Performance and Minimum Scores

As stated before, in the PST you must achieve minimum scores. Here we provide the minimum scores, plus points of performance—what is required of you in terms of how to execute each exercise. Use the minimum scores as a guide, and work to achieve higher scores—it cannot be said often enough—the higher your scores during boot camp, PRE-BUD/S and BUD/S, the better your chances. Squeaking by with the minimum is not a good plan.

Points of Performance

⌃ **U.S. Navy SEALs and special warfare combatant-craft crewmen (SWCC) recruiters attend the 18th annual Hispanic Track and Field Games. The U.S. Navy-sponsored event attracted 7,000 high school graduates, parents and spectators.**

Credit: U.S. Navy Mass Communication Specialist 2nd Class Meranda Keller.

Swim

You must perform the test using either a Side Stroke or a Breast Stroke.

Push-ups

Must be performed with a straight back and feet and hands in contact with the deck at all times. No slouching! Proper form must be strictly maintained.

Sit-ups

Sit on the deck with your knees bent approximately 90 degrees. Cross your arms in front of you with fingertips touching your shoulders. Maintain full range of motion.

Pull-ups

Grip pull-up bar with *your palms facing away from you*. Hands are shoulder width apart. Do not swing, kick, kip, or bicycle your legs to assist. Make sure you go all the way up (your chin must be above the bar) then all the way down.

> Side note: kipping during pull-ups is considered cheating by some, and others view kipping as a technique that allows for increasing power output.

SEAL Physical Screening Test (PST) Minimum Requirements

PST	Minimums
SWIM 500 YDS. Side stroke/ breast stroke	12:30 min
REST 10 MIN.	
PUSH-UPS (within 2 min)	42
Rest 2 minutes	
SIT-UPS (within 2 min)	50
Rest 2 minutes	
PULL-UPS (no time limit)	6
Rest 10 minutes	
1.5 MILE RUN	11:00 min

SEAL *Competitive* Physical Screening Test Scores

PST	Score
SWIM 500 YDS. Side stroke/breast stroke	10:30 min
REST 10 MIN.	
PUSH-UPS (within 2 min)	79
Rest 2 minutes	
SIT-UPS (within 2 min)	79
Rest 2 minutes	
PULL-UPS (no time limit)	11
Rest 10 minutes	
1.5 MILE RUN	10:20 min

The ASVAB Test

(See Appendix 1 on page 137 for sample questions.)

As said earlier, physical aptitude is only part of the equation when becoming a SEAL candidate. If you are serious about becoming a Navy SEAL—then you are going to have to pass the Armed Services Vocational Aptitude Battery, or ASVAB. This is a test of basic knowledge such as math, verbal and writing skills, and vocabulary. It measures your strengths, weaknesses, and success quotient, and also provides you with information on various military and civilian occupations.

Unless you are entering an officer's training program, you are required to take the test no matter what branch of the military you plan to join. The ASVAB does not only screen you for entrance into the military, it is also used for possible job placement. This is why you must take this test very seriously and not just try to breeze through it thinking all you have to do is pass it; that's not how it works! You'll want to focus on your strongest skills and getting the best scores possible—higher scores mean a wider variety of choices: better jobs, higher salaries, and more opportunities.

There are three versions of the ASVAB exam. The CAT-ASVAB (computer adaptive test) can only be taken at Military Entrance Processing Stations (MEPS) when you enlist. The exam is timed and it will not be possible to go back and check or change answers after you've submitted them. The CAT-ASVAB is in ten parts and includes topics such as arithmetic, verbal skills, general science, and mechanical knowledge.

Start practicing now! For a sample and full-length practice tests visit www.military.com/join-armed-forces/asvab

The MET (Mobile Examination Test) Site ASVAB is only administered for enlistment into one of the branches of the military after you've been referred by a recruitment officer. It is in eight parts and is taken with pencil and paper. It is still a timed test, but with the MET–ASVAB you can check and change your answers. There are no penalties for wrong answers with this test, so answer all of the questions with your best guess if you don't know the answer—this will increase your chances of scoring well.

The Student ASVAB is provided to high school students to help them assess their skills, job prospects, possible military careers, or college majors. It is, for the most part, the same as the MET ASVAB, only students aren't necessarily taking the test to enter the military, but to get an idea of their strengths and weaknesses while planning for higher education and career goals.

≈ **Aboard USS *Kearsarge* (LHD 3) in the Arabian Sea a Marine Aviation Maintenance Mechanic visually checks rotor locks on a CH-53E Sea Stallion helicopter prior to flight.**
Credit: U.S. Navy photo by Photographer's Mate 2nd Class Alicia Tasz.

The subtests that make up the ASVAB focus on basic knowledge of science, math, writing and vocabulary, an understanding of structural development and mechanics, auto function and repair, a knowledge of electric currents, electronic systems, and circuits. These are all skills and knowledge that are necessary to have at various times and in various sectors of your military service. You are scored on your ability to answer the questions correctly, complete the test on time and on how many questions you answered.

⚑ **An Aviation Electronics Technician Airman conducts a temperature controller test set for a cabin sensor aboard USS *Harry S. Truman*.**
Credit: U.S. Navy photo by Photographer's Mate 3rd Class Danny Ewing, Jr.

ASVAB Subtests

Subtest	Questions	Minutes	Description
General Science (GS)	25	11	Measures knowledge of physical and biological sciences
Arithmetic Reasoning (AR)	30	36	Measures ability to solve arithmetic word problems
Word Knowledge (WK)	25	11	Measures ability to select the correct meaning of words presented in context, and identify synonyms
Paragraph Comprehension (PC)	15	13	Measures ability to obtain information from written material
Auto and Shop Information (AS)	25	11	Measures knowledge of automobiles, tools, and shop terminology and practices
Mathematics Knowledge (MK)	35	11	Measures knowledge of high school mathematics principles
Mechanical Comprehension (MC)	25	19	Measures knowledge of mechanical and physical principles, and ability to visualize how illustrated objects work
Electronics Information (EI)	20	9	Tests knowledge of electricity and electronics

★★★★★

Each branch of the military requires a different minimum score for its members. To become a SEAL candidate your GS (General Science) + MC (Mechanical Comprehension) + EI (Electronic Information) must equal 165, or your VE (Verbal Expression, which is based on your WK [Word Knowledge] and PC [Paragraph Comprehension] results) + MK (Mathematics Knowledge) + MC (Mechanical Comprehension) + CS (Coding Speed) must equal 220. Please note: It is crucial that you score high on the ASVAB since you cannot get a waiver on the required ASVAB score for Navy SEAL candidacy.

This does not mean waivers are not available for other Navy jobs or other branches of the service. Check with your recruiting officer for information on waivers.

C-SORT: Testing Mental Toughness and Resilience

The Computerized-Special Operations Resilience Test, or C-SORT, assesses a prospective SEAL candidate's mental toughness. The test covers three areas:

- Performance strategies
- Psychological resilience
- Personality traits

Performance strategies test for skills such as a person's goal-setting and emotional control.

Psychological resilience assesses an individual's acceptance of life situations, and the ability to deal with cognitive challenges and threats.

The scores on the sections are combined into one score based on a scale of one to four, called a band score. A band score of four indicates high mental resilience, a one, the lowest level. As a prospective SEAL candidate, you can only take the C-SORT one time.

Your C-SORT band score will be combined with your run and swim time to determine eligibility for the SEAL program. If your scores are low, you will not be considered for SEAL candidacy. But all is not completely lost. Even though you cannot retake the C-SORT, if you improve your PST score, specifically run and swim times, and retake the Delayed Entry qualifying PST, your combined score might improve enough to be eligible for a SEAL contract.

Once you have fulfilled the minimum standards and have your SEAL contract, you will start training with your SEAL mentor during the Delayed Entry Program (DEP) phase. When you reach optimum levels, you are ready to go to boot camp at the Recruit Training Center at Great Lakes, Illinois. But before you go, your PST score will be ranked against national scores and only those with the best scores—called optimum range plus—will be chosen for that boot camp cycle. This is called the Navy Spec War Draft—not to be confused with Selective Service.

If you are not selected, you can take more time to train, or make the decision that you have skills that might work better in another vocation or branch of the military.

⌃ **U.S. Navy recruits stand at attention following the successful completion of Battle Stations, the final portion of Navy recruit training, at Recruit Training Command Great Lakes, Illinois.**

Credit: DOD photo by Chief Petty Officer Johnny Bivera, U.S. Navy.

After You Complete Boot Camp

Once you have finished boot camp you will attend PRE-BUD/S, a seven- to nine-week Naval Special Warfare Preparatory School in Great Lakes, Illinois, designed to physically prepare you for BUD/S. If you had any difficulty at all in boot camp, you need to reconsider applying to BUD/S.

Essentially, PRE-BUD/S will help bridge the enormous physical training gap between regular boot camp and BUD/S. During your weeks at PRE-BUD/S you will be stressed beyond your limits. The Navy wants to make sure that you are the best and most worthy to join its elite fighting force, but it also wants to give you a chance to improve your physical abilities. Do yourself a favor and be prepared: Understand fully what becoming a SEAL means and what you're volunteering for *before* you go to PRE-BUD/S.

Make no mistake, PRE-BUD/S is another weeding out process, just like the minimum requirements, PST, the ASVAB and the C-SORT. Making it through PRE-BUD/S at optimum level is just as important as in any other phase. You must never take these preliminary tests and phases lightly if you want to stand a chance at making it to BUD/S. I highly recommend that while in boot camp you spend a couple hours a day doing SEAL physical training (PT). You will not stay in shape if all you do is the basic boot camp PT.

4

TRAINING, EDUCATION, AND PLACEMENT

So far we've discussed some of the hurdles you will have to clear just to get accepted into BUD/S, and you have some knowledge of the history of the SEAL Teams and the kinds of missions the teams are assigned. To be a SEAL is to be a part of a team—it cannot be emphasized enough. But what of your own personal contribution: that skill set or expertise you may possess and excel at that you can bring to the teams?

As a Navy SEAL you will have plenty of opportunities for advanced training and to gain expertise in a wide variety of military disciplines. During your career as a SEAL you will attend special schools and earn qualifications in a number of skills and specialties, such as Advanced Demolitions and Breaching,

Parachute Rigger, Rappel Master, Fast Rope Master, Freefall Jumpmaster, Diving Supervisor in Open and Closed-Circuit Diving, Lead Climber, Medic, Survival Instructor, and Sniper. Many SEALs have the opportunity to acquire

⌃ A Navy SEAL Sniper is assisted by his spotter, who calls wind direction and determines where the rounds impact through the spotting scope.

most of these special-ties throughout their careers. For example, all SEALs learn basic demolitions but many go on to become the team demolition spe-cialist and receive

advanced training in breaching and other techniques. You will train in urban, jungle, mountain, and desert warfare, and a few of you may be fortunate enough to even train in arctic terrain.

An enlisted SEAL will be assigned a SEAL team or Special Delivery Vehicle (SDV) team

Note: a military "staff" is a group of officers and enlisted personnel that is responsible for the administrative, operational, and logistical needs of its unit.

SEAL OFFICER CAREER TRACK

Officers and enlisted men go through BUD/S and SQT (SEAL Qualification Training) together, as they train, operate, live, and fight together. The difference is that officers eventually move out of the field and operational roles to a desk—to the planning and strategizing level. As an officer you will be an assistant platoon commander in an assistant officer in charge billet, and later a platoon commander in an officer in charge billet.

Officers also have their choice of disciplines in which they can train, but do not generally train in enlisted specialties such as Special Demolitions, Combat Medicine, Parachute Rigging, and so on. By an officer's third tour he will be assigned shore duty on a staff, in language school, or at the Naval Postgraduate School in Monterey, California, or may attend the U.S. Military War College in Newport, Rhode Island.

The officer's career will include operational command and staff tours, with the opportunity to become a commanding officer of a SEAL team, to command a Naval Special Warfare Group or Naval Special Warfare Development Group, or to be, among other things, a staff officer at the Pentagon, a member of a senior Navy staff, a deputy commander of a theater, or a special operations commander.

and remain on this tour of duty for three to five years. You can request the location of your choice—east or west coast, and the Navy will make an effort to fulfill that request provided it jibes with the needs of the military.

Further on in your career—eight to ten years down the road—you can qualify to be a BUD/S instructor or to serve in other select Navy or joint service billets. For enlisted personnel, most of your career will be spent conducting training and operations. Without exception, as a SEAL you will have the opportunity to advance in enlisted and/or officer leadership.

The needs of the military, your skills and performance, available assignments and billets, and the timing of rotation (the changing of assignments and tours) will always factor into your career trajectory.

★★★★★

So you've gotten through basic training, or you've done your requisite tours, and there is nowhere else you'd rather be but with the SEAL teams. So now you *must* get accepted to BUD/S and if you make it through BUD/S, to Parachute Jump School, and then on to SEAL Qualification Training (SQT).

Before all that you must attend PRE-BUD/S—SEAL Prep, or more formally, Naval

- 1000-meter swim—with fins (22 minutes or under)

- Push-ups: at least 70 (Two-minute time limit)

- Pull-ups: at least 10 (No time limit)

- Curl-ups: at least 60 (Two-minute time limit)

- Four-mile run: with shoes + pants (31 minutes or under)

Special Warfare Preparatory School (NSW prep)—mentioned in the previous chapter.

If you do not pass the final tests at NSW prep, you will be taken off the SEAL track and reclassified to other billets in the Navy, it's as simple as that. If you can't pass muster, you can't possibly expect to make it through the much more brutal BUD/S training.

The prep school begins and ends with a physical screening test (see sidebar above for minimum standards to pass the test).

Along with physical training, which will include swimming, basic underwater skills, strength and conditioning, running, and calisthenics, the potential candidate will also be taught nutrition, exercise science, injury prevention, rest and recovery, keys to mental toughness, basic military training, and given instruction in professional development. Other topics taught are military heritage; military rights and responsibilities; morale, welfare and

THE SEAL ETHOS

In times of war or uncertainty there is a special breed of warrior ready to answer our Nation's call. A common man with uncommon desire to succeed. Forged by adversity, he stands alongside America's finest special operations forces to serve his country, the American people, and protect their way of life. I am that man.

My Trident is a symbol of honor and heritage. Bestowed upon me by the heroes that have gone before, it embodies the trust of those I have sworn to protect. By wearing the Trident I accept the responsibility of my chosen profession and way of life. It is a privilege that I must earn every day.

My loyalty to Country and Team is beyond reproach. I humbly serve as a guardian to my fellow Americans always ready to defend those who are unable to defend themselves. I do not advertise the nature of my work, nor seek recognition for my actions. I voluntarily accept the inherent hazards of my profession, placing the welfare and security of others before my own.

I serve with honor on and off the battlefield. The ability to control my emotions and my actions, regardless of circumstance, sets me apart from other men. Uncompromising integrity is my standard. My character and honor are steadfast. My word is my bond.

We expect to lead and be led. In the absence of orders I will take charge, lead my teammates and accomplish the mission. I lead by example in all situations.

I will never quit. I persevere and thrive on adversity. My Nation expects me to be physically harder and mentally stronger than my enemies. If knocked down, I will get back up, every time. I will draw on every remaining ounce of strength to protect my teammates and to accomplish our mission. I am never out of the fight.

We demand discipline. We expect innovation. The lives of my teammates and the success of our mission depend on me—my technical skill, tactical proficiency, and attention to detail. My training is never complete.

We train for war and fight to win. I stand ready to bring the full spectrum of combat power to bear in order to achieve my mission and the goals established by my country. The execution of my duties will be swift and violent when required yet guided by the very principles that I serve to defend.

Brave men have fought and died building the proud tradition and feared reputation that I am bound to uphold. In the worst of conditions, the legacy of my teammates steadies my resolve and silently guides my every deed. I will not fail.

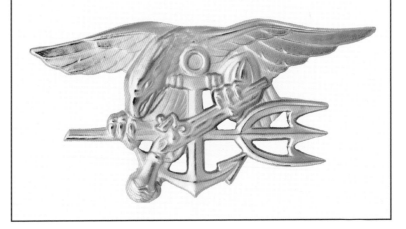

recreation; sexual assault, harassment, fraternization, and discrimination. Perhaps most importantly candidates will be taught the SEAL ethos, the guiding belief that ends with the words "I will not fail."

SO YOU'VE MADE IT TO BUD/S (6+ Months)

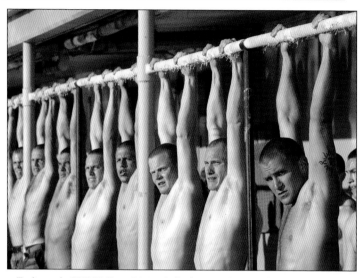

⌃ Trainees in BUD/S Class 244 wait for instructions as they prepare to execute pull-ups as part of their morning exercise routine at the Naval Amphibious Base Coronado, San Diego, California.

You finally made it to BUD/S. Now you will be devoting six-plus months of your life to training at the Naval Special Warfare Training Center in Coronado, California. The first five weeks will be indoctrination and pretraining, and then you will go through the three phases of BUD/S.

⌃ A BUD/S trainee wades ashore on San Clemente Island during an over-the-beach exercise.

Credit: U.S. Navy photo by Mass Communication Specialist 2nd Class Kyle D. Gahlau.

≈ A BUD/S trainee moves through the weaver during an obstacle course evolution in the first phase of training at Naval Amphibious Base Coronado.

Credit: U.S. Navy photo by Mass Communication Specialist 2nd Class Kyle D. Gahlau/Released.

≈ BUD/S trainees prepare to participate in interval swim training in San Diego Bay. The swim is part of the basic orientation phase of BUD/S.

Credit: U.S. Navy photo by Mass Communication Specialist 2nd Class Trevor Welsh.

⌃ **First-phase BUD/S trainees navigate small inflatable boats (IBS) through the surf during a maritime navigation training evolution.**

Credit: U.S. Navy photo by Mass Communication Specialist 2nd Class Shauntae Hinkle-Lymas.

⌃ **A BUD/S trainee participates in interval swim training in San Diego Bay.**

Credit: U.S. Navy photo by Mass Communication Specialist 2nd Class Trevor Welsh.

BUD/S is never easy, but the eight weeks of basic conditioning that you will go through in the first phase—among other things, running, swimming, obstacle courses, rope climbs, paddling, lots of PT with "Hell Week" at the halfway point—will be the toughest part of BUD/S.

Hell Week will test your physical endurance, teamwork, and mental resilience. You will be cold and wet. You will be sleep deprived and fatigued. You may even hallucinate. You will be in discomfort most of the time. You might even wonder why you ever decided to try to be a SEAL and question your own sanity.

More than likely two out of three of your classmates will quit.

But a few of you will make it through despite everything the BUD/S instructors throw at you. All you need to do is "Welcome the Pain."

≫ **Trainees in BUD/S Class 279 participate in a surf passage exercise during the first phase of training in Coronado, California. Surf passage is one of many physically strenuous exercises that BUD/S classes take part in during first phase.**

Credit: U.S. Navy photo by Lt. Frederick Martin/Released.

⌃ **Trainees from BUD/S class 286 participate in a surf passage training exercise.**
Credit: U.S. Navy photo by Mass Communication Specialist 2nd Class Kyle D. Gahlau.

⌃ **BUD/S trainees rush to get their inflatable boats (IBS) to the finish line in a surf passage evolution during Hell Week.**
Credit: U.S. Navy photo by Mass Communication Specialist 2nd Class Marcos T. Hernandez.

⚐ A BUD/S instructor is about to show a trainee from BUD/S Class 244 just how difficult it can be to rescue a drowning victim when the "victim" comes at his rescuer with a vengeance during lifesaving training.

Credit: U.S. Navy photo by Photographer's Mate 3rd Class John DeCoursey.

« BUD/S trainees swim 100 meters with bound hands and feet as part of their first-phase swimming test.

Credit: U.S. Navy photo by Mass Communication Specialist 2nd Class Shauntae Hinkle-Lymas.

You now are ready to prepare for second phase.

The second phase lasts for eight weeks and during that time you will be taught open and closed circuit diving and you will learn the basic skills of becoming a Combat Diver.

Phase three is nine weeks of land warfare.

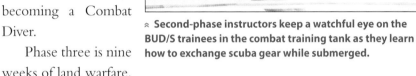

⌃ **Second-phase instructors keep a watchful eye on the BUD/S trainees in the combat training tank as they learn how to exchange scuba gear while submerged.**

Most likely, the only thing that might stop you during these phases is academic problems

« **Trainees from BUD/S Class 287 participate in night gear exchange during the second phase of training at Naval Amphibious Base Coronado. During this session, two trainees will enter the water and exchange dive gear with masks that have been completely blackened.**

Credit: U.S. Navy photo by Mass Communication Specialist 2nd Class Kyle D. Gahlau.

⚹ A second-phase BUD/S trainee checks his underwater breathing apparatus for ventilation and safety before participating in dive training in Coronado Bay. Diving is an eight-week course during the second phase of training.

Credit: U.S. Navy photo by Mass Communication Specialist 2nd Class Shauntae Hinkle-Lymas.

⚹ A U.S. Navy BUD/S trainee signals during a simulated dive casualty drill during training at Naval Amphibious Base Coronado, California.

Credit: U.S. Navy photo by Petty Officer 2nd Class Dominique M. Canales.

≈ A third-phase trainee in BUD/S detonates a SEAL Standard Charge on San Clemente Island, California.

Credit: U.S. Navy photo by Mass Communication Specialist 2nd Class Blake Midnight.

≈ Trainees in third phase of BUD/S rappel down a 60-foot tower in Coronado. For those trainees who go on to become Navy SEALs, this marks the first of many rappels to come in their careers.

Credit: U.S. Navy photo by Mass Communication Specialist 2nd Class Blake Midnight.

⌃ A third-phase trainee in BUD/S fires his rifle down range during a night live-fire exercise. The third phase of BUD/S training focuses on land warfare and includes training in pistol, rifle, demolitions and tactical movement.

Credit: U.S. Navy photo by Mass Communication Specialist 2nd Class Blake R. Midnight.

« A BUD/S trainee reads his compass during basic land navigation training. Basic land navigation teaches candidates how to read a map, plot coordinates, and navigate over various types of terrain.

Credit: U.S. Navy photo by Mass Communication Specialist 2nd Class Dominique M. Lasco.

» **A group of BUD/S trainees plot coordinates together during basic land navigation training.**

Credit: U.S. Navy photo by Mass Communication Specialist 2nd Class Dominique M. Lasco.

during diving training or weapons competency issues during land warfare training.

The final eight weeks are devoted to learning core tasks: mission planning, operations, and tactic, techniques and procedures. Once BUD/S is

⌃ **The desert near Niland, California, offers a perfect stage for live-fire exercise, day and night.**

Credit: Photo by Photographer's Mate 1st Class (AW) Shane T. McCoy.

complete, candidates attend three weeks of basic parachute training at Fort Benning, Georgia.

After you complete BUD/S you have earned the right to wear the coveted SEAL Trident and you go on to SEAL Qualification Training (SQT).

BUD/S Phases

Phase 1—Physical Conditioning (8 weeks)

- Running in the sand

- Swimming—up to 2 miles with fins in the ocean

- Calisthenics

- Timed Obstacle Course

- Four-mile timed runs in boots

- Small boat seamanship

- Hydrographic surveys and creating charts

- Hell Week—Week 4 of Phase 1 (five and a half days of continuous training, four hours sleep, total)

- Swimming

- Running

- Enduring cold, wet, and exhaustion

- Rock Portage in Combat Rubber Raiding Crafts (CRRC)

- TEAMWORK!

Phase 2—Diving (8 weeks)

- Step up intensity of the physical training

- Focus on Combat Diving

- Open-Circuit (compressed air) SCUBA

- Closed-Circuit (100% oxygen) SCUBA

- Long-distance underwater dives

- Mission-focused combat swimming and diving techniques

Phase 3—Land Warfare (9 weeks)

- Increasingly strenuous physical training

- Weapons training

- Demolitions

- Small unit tactics

- Patrolling techniques

- Rappelling and fast-rope operations

- Marksmanship

HELL WEEK

⌃ **During a Hell Week surf drill evolution, a Navy SEAL instructor assists trainees from BUD/S class.**

Credit: U.S. Navy photo by Photographer's Mate 2nd Class Eric S. Logsdon.

The fourth week of Phase I is known as Hell Week. It is during this intense five-and-a-half-day period that most candidates withdraw (known as Drop on Request, or DOR), unable to contend with the

⌃ **BUD/S trainees perform Log PT during Hell Week. During Hell Week a special bell is ever ready; should a student decide he no longer wishes to continue with the training, he can ring the bell.**

Credit: U.S. Navy photo by Mass Communication Specialist 2nd Class Marcos T. Hernandez.

grueling and nonstop regimen of mental and physical workouts. It is designed specifically for this purpose. The Navy's experience has been that the most common characteristic of the SEAL candidate who successfully completes BUD/S is not physical prowess, but mental hardiness and, most importantly, personal determination. Those candidates who approach Hell Week with a mentality of non-failure are more likely to survive the five-and-a-half-day battering.

Hell Week is an extension of Phase I's incremental training program (referred to as "evolutions"), but to the extreme. Over the course of these five and a half days, SEAL candidates are tasked with completing the same strength, endurance, and teamwork exercises from the past three weeks: long slow distance running and swimming; continuous high intensity running and swimming; intervals alternating between low and high intensity workouts; and team based endurance exercises.

In addition to the continuous exercises that SEAL candidates must complete, they must also endure the relentless pressure from their BUD/S trainers. These trainers will prod and question each candidate's commitment to the SEAL program, attempting to sow seeds

≈ **BUD/S trainees brace for impact during whistle drills during a combat scenario-driven field training exercise during Hell Week in Coronado, California.**

Credit: Navy photo by Mass Communication Specialist 2nd Class Marcos T. Hernandez.

of doubt and, in so doing, weed out the candidates who are not pre-pared to embody the SEAL ethos. Throughout this all, the candidates are only allowed a maximum of four hours of sleep while being tasked with running nearly 200 miles and exercising for 20 hours a day.

Only those candidates who successfully endure the mental and physical duress, exposure to extreme temperatures and conditions, and constant berating of their trainers are allowed to continue on through Phase I. Of the more than 70 percent of candidates who with-draw from the SEAL program, 57 percent occur during Phase I and its redoubtable Hell Week.

» **BUD/S instructor congratulates trainees in Basic Underwater Demolition/SEAL (BUD/S) Class 290 upon their completion of Hell Week at Naval Amphibious Base Coronado.**

Credit: U.S. Navy photo by Mass Communication Specialist 2nd Class Kyle D. Gahlau.

≈ **BUD/S trainees who have completed through Hell Week raise a cheer.**
Credit: Photo by Photographer's Mate 1st Class (AW) Shane T. McCoy.

« A cold-weather training instructor monitors Navy SEALs as they spend five minutes in near-freezing water as part of SEAL Qualification Training (SQT) at Kodiak, Alaska.

⌃ A U.S. Navy SEAL in Naval Special Warfare SEAL Qualification Training rappels from a 60-foot tower at the Naval Special Warfare Center in Coronado, California.

⋩ SEAL Qualification Training students from Class 268 perform buddy carries between stations during a shooting drill at Camp Pendleton in California, 2008.

Credit: U.S. Navy photo by Mass Communication Specialist 2nd Class Michelle Kapica.

⋩ U.S. Navy SEAL Qualification Training students ride an inflatable boat (IBS) in San Diego Bay after plotting a course on a map during their 12 days of maritime operations training.

⋩ A Special Warfare Combatant-craft Crewman (SWCC) from Special Boat Team 12 treats an injured teammate during a casualty assistance and evacuation scenario at Naval Special Warfare Center at Camp Pendleton, California, August 2008.

Credit: U.S. Navy photo by Mass Communication Specialist 2nd Class Michelle Kapica/ Released.

« SEAL Qualification Training students parachute out of a C-130 Hercules military transport aircraft during a routine training exercise over San Diego Bay.

Credit: U.S. Navy photo by Mass Communication Specialist 2nd Class Dominique M. Lasco.

⌃ Students in SEAL Qualification Training navigate the surf off the coast of Coronado, California, during a maritime operations training exercise.

⌃ A student in SEAL Qualification Training Class 279 conducts immediate action drills to learn how to react as part of a team during enemy contacts.

Credit: U.S. Navy photo by Mass Communication Specialist 3rd Class Blake Midnight.

⚹ SEAL Qualification Training students take aim during a 36-round shooting test ranging from 100, 200, and 300 yards at Camp Pendleton.

Credit: U.S. Navy photo by Mass Communication Specialist 2nd Class Michelle Kapica.

⚹ SEAL students scan the room for possible threats as part of a SEAL qualification training exercise.

Credit: U.S. Navy photo by Mass Communication Specialist 2nd Class Christopher Menzie.

SEAL QUALIFICATION TRAINING (SQT) (26 weeks)

Completing BUD/S only *qualifies* you for the right to earn your SEAL trident. It takes some five years to train a SEAL to a high level of competency in all skills and mission areas. You've made your family and the Navy proud, but now it's time to report in to your assigned SEAL team and attend SEAL Qualification Training (SQT).

The 26-week course is designed to hone individual skills as much as it will take you from the basic elementary level of BUD/S to a more advanced degree of tactical training. SQT is designed to provide students with the core tactical knowledge they will need to join a SEAL platoon. More importantly, they learn how to operate as a team.

SQT begins with classroom work in mission planning and intelligence gathering and reporting. You then begin a series of "blocks" of training that will train you in the major skills needed to conduct SEAL missions: hydrographic reconnaissance, field medicine, communications, combat swimming, air skills, maritime operations, land warfare, and submarine lock-in/lock-out.

During Air Week you will conduct day and night static line jumps and a water jump accompanied by a "rubber duck" (an inflatable Zodiac boat with motor and gear dropped out of the back of an aircraft under canopy), followed by six or so SEALs to chase it to the water. You will learn fast-rope techniques (sliding down a nylon rope for 80 feet with only your gloved hands to stop) and rappelling techniques. You will be introduced to the "Special Insertion/Extraction" (SPIE) rig, a vehicle that can extract six to eight SEALs from an area too rugged or dense to land a helicopter.

Combat swimmer training will have you conducting more than 25 day and night compass dives, beginning with the basics and progressing to full mission profiles. Land warfare training, usually held at the Naval Special Warfare training facility at

Niland, California, or at Camp A. P. Hill near Bowling Green, Virginia, is a three-week course that sharpens SEAL skills in patrolling, stalking, weaponry and military demolition, and creating improvised booby traps. You will take part in live fire immediate action drills, where the team fires and maneuvers in well-choreographed sequences, and learn how to perform in "fog of battle" scenarios—drills conducted often at night, with reactive targets, pop flares, and smoke grenades that cause confusion and chaos.

Before graduating, candidates also attend SERE (Survival, Evasions, Resistance, Escape) training, which is exactly like it sounds, complete with simulated POW camps and "torture" scenarios.

★★★★★

SERE Training

Since I have been traveling to war zones and places of "unrest" for the majority of my life, I have no choice but to continually maintain, sharpen and fine-tune my own survival mindset, my "survivor's awareness." I see no reason to turn this off once I get back stateside because, as we all know, we have threats roaming throughout our country and these threats only seem to multiply year after year. I am not paranoid, just aware of my own surroundings and confident of my training and ability to react to emerging threats.

SEALs are confronted with dangerous situations—combat, operating in hostile environments, hazardous training, and so on—all the time. A toast you will frequently hear from SEALs is "Here is to cheating death once again!" SEALs are very successful in what they do because they possess a "combat mindset" and are armed with a great deal of collective war fighting experience and the knowledge. As one of the SEAL team's earliest and most renowned weapons instructors, Jeff Cooper, once stated, "The most important means of survival in a lethal confrontation is neither the weapon nor martial arts skills. The primary tool is the combat mindset."

SEALs also possess a survival mindset, and are armed with the knowledge of the appropriate survival techniques.

From the time when the first SEAL teams were commissioned, to the present, they have distinguished themselves as being individually reliable, collectively disciplined and highly skilled. Because of the dangers inherent in what they do, SEALs go through what is considered by military experts to be the toughest training in the world—Basic Underwater Demolition SEAL Training (BUD/S).

During BUD/S the trainees develop a "combat mindset" as well as "survival mindset." However, they really are not exposed to "survival training." Survival training typically begins at the Navy Survival Evasion Resistance and Escape (SERE) school.

As a young Navy SEAL recently graduated from BUD/S, class 120, I was told by a Vietnam era SEAL that if a SEAL was captured during wartime there was a good chance that he would be beheaded or skinned alive. I immediately volunteered to attend the SERE course conducted at Warner Springs, California.

I took my survival training very seriously. We were given some basic training in land navigation, poisonous and edible plants, poisonous and edible animals and insects, water procurement, fire making, building shelters, and evasion and escape techniques.

After these training sessions were completed we were dropped off in the desert and were on our own for food and water procurement while navigating to a safe area and trying to avoid contact with the "enemy."

I got very lucky. We were all really hungry and were looking for anything to eat. We drank from the Prickly Pear cactus and looked for edible plants. I happened to see a small rabbit running under a bush. I threw my KA-BAR knife at it, just hoping to hit it. To my surprise the knife pinned the rabbit's neck into the ground. I skinned it and made rabbit stew for the team. There were over 20 of us in the course. After the rabbit was skinned we realized just how small it was and how little meat it had on it. Along with the edible plants and rabbit the stew ended up tasting very much like hot water. We needed a lot more food. I realized then how important it was to receive proper training in survival and water and food procurement.

SERE school is designed so eventually all participants are captured. When I was captured, I was tie-tied, blindfolded, and put into an "enemy" Jeep. The instructors, outfitted in realistic communist-style clothing all stayed in "role" as they barked orders with a cold war Russian accent.

Because I considered this training to be so important, I played it for real. My captor stepped out of the Jeep leaving me alone with his only means of communication, a PRC-77 radio. I was able to get my bound hands in front of my body, picked up the radio and threw it under the vehicle. I also had a knife and a lighter hidden in my boots.

We were eventually driven to the POW training camp. A fenced-in area, with numerous enemy guards interrogating other "prisoners," slamming them into walls, and humiliating them by having them stand naked while being drilled with questions and slapped in the face. The prison guards and interrogators had a wide variety of very effective techniques that they used to obtain information from their prisoners.

In soft cell, you were often called into a warm office, with either a pretty woman or a gentleman guard. They understood how mean the other guards were and even offered you warm coffee, snacks, and sometimes warm, dry clothing. It was a stark contrast from the hard cell interrogations.

They were successful in that many of their captives broke in hard cell and surprisingly, many also broke and gave information to the kinder-gentler captors in these soft cell sessions.

Military personnel do their best to abide by their chain of command while in captivity. The highest ranking captive was typically in charge of the group. But, usually unbeknownst to the enemy captive, one of the prisoners would be put in charge of managing prisoner escapes. The most senior captive often would select a fellow captive to be the "covert leader of escape." It was this person's job to do whatever he could to free his fellow captives.

During my SERE training I was the only SEAL. The other students were pilots or air crew personnel. So I was selected to be the covert leader of escape. I probably went a bit overboard with this responsibility. I managed to sneak out of my three-foot by three-foot cell, and after numerous attempts, managed to smash all of the lights that lit the inner camp. This took many hours to accomplish, sneaking in and out of my cell and throwing the rocks only when the guards were out of sight. I also snuck into all of the other cells and took the potato sacks out. While in our cells, we were to sit cross-legged on the floor, left foot over the right foot, sack over our head. Our heads were cocked to the side since the boxes were built a bit too small for comfort.

When the guards walked past your cell they would bang on the roof and you were to report "war pig #4 here, Sir!" Throughout the day and night, you would hear this roll call shouted out from each individual cell. The trick was to try to leave your cell after roll call and get back before the next roll call. During one of my trips taking the sandbags over to the fence line to conceal them, war pig #8 answered, as did 7, 6, 5, but war pig #4 was absent. I snuck into the cell of a fellow captive. It was very crowded and he was scared to death that he would be "tortured" if the guards found out he was aiding another prisoner who was breaking camp rules.

As the course progressed I was surprised at how many of my fellow captives actually broke down. There were some who ended up believing that they were actually in an enemy POW camp. Some broke down and cried. The training was realistic and effective.

One bitter cold morning we were forced to all crawl on our hands and knees and had to draw a hammer and sickle into the dirt. Our bodies were tired, thirsty, and broken down, but for the most part everyone did their best and tried to stay out of the focus of the guards. One of the more hostile guards was announcing over the loud speaker how terrible the United States is, how rotten our country's leaders are, and how fortunate we were because we were being given an opportunity to just simply give up the information they were asking for. It all seemed so easy—for some. A number of the captives gave up—giving all the information that they were asked for and were rewarded with warm drinks, warm food, and warm clothing. The rest of us who were naked or barely clothed and crawling in the dirt were forced to listen to the soldier spieling out his anti-U.S. propaganda.

They threatened us with the infamous water board. They described how dreadful the feeling of drowning is and how after they strap our feet to the foot of the board, which was about four feet off the ground, and our heads, near ground level, with a cold rag thrown over our mouths, we would choke and cough as they poured water on the rag and into our mouths and noses.

As with most things in life, it was not nearly as bad as people make it out to be. I ended up having the distinction of spending over thirty minutes on the water board. Initially the guard who was spieling all of the propaganda asked over the mega phone, "Who is arranging your escapes? We know you have a covert leader of escapes who has somehow managed to get seven of you pigs out of our prison. Unless you tell us who your covert escape leader is, we will put each and every one of you on the water board and will enjoy watching you choke on your own vomit." He walked into the dirt hammer and sickle and grabbed one of the smallest captives and placed him next to the water board. He said, to the scared lieutenant, "You have one more chance—who is getting you all out of this prison?"

I was impressed because the lieutenant didn't give in. But as soon as they lifted him and began strapping him to the water board, he broke down and shouted, "It's the SEAL. The SEAL is getting the other captives out of camp. He cut a hole in the fence and is sneaking people out whenever he gets a chance."

Well, the guards came over and picked me up from the dirt and strapped me to the water board. It was uncomfortable, but I soon realized that the more I coughed and choked, the less water they poured into my mouth. Between the water treatments they would interrogate me. But as we were taught in training, we do not give up anything other than name, rank, and social security number. And by no means would I admit to being a SEAL or as they called me a "baby killer."

We could only give additional information to the interrogators if we were in fear of losing life or limb. It was more of an art than it was a science in how to answer and not answer their questions. I was proud that I took the punishment and did not divulge any information. It was a victory and I beat them.

After a while, we were all thrown back into our cells. That afternoon, two guards came and pulled me out of my cell and marched me to the commandant's office. I was in trouble for something. They pushed me up against the wall and directed me to stand at attention. The commandant stepped into the office, stood directly in front of me, looked directly into my eyes, and said, "Is it you getting other captives to escape?"

"No, Sir."

"Where is my PRC 77?"

"I do not know, Sir."

"Where are the other potato sacks?"

"I do not know, Sir."

"Did you smash our lights?"

"No, Sir."

"I need to know the truth."

All I gave was my name, rank, and social security number. The commandant was becoming increasingly irritated. He was out of role, and I stayed in role. He really needed to know these answers as the director of the SERE course. I needed to stay in role and take my training seriously. Finally he shook my hand and told me, "You have been selected as the class Honor Graduate. I will go to war with you anytime, but I need to know the answers to my questions."

The course was over. I told him what he needed to know, and he told me that he would like for me to raise the American flag to signify the end of the course. It was an honor, and I was so proud to have completed this course.

The commandant's name was Captain Gaither. Captain Gaither is an American hero who survived captivity in Viet Nam.

From *The U.S. Navy SEAL Survival Handbook*

You survived BUD/S and SQT. You now stand in your dress Navy uniform, in the presence of family and senior SEAL leaders—commanding officers and senior enlisted advisors of Naval Special Warfare Groups and SEAL teams. You are a Navy SEAL—one of the few, one of the elite, and you are presented with the coveted Navy SEAL trident, one of the most recognizable insignias in the U.S. military. In the teams the trident is also called the "Budweiser."

5

THE JOURNEY BEGINS

When the trident is pinned to your uniform, and you are filled with pride in your accomplishment—you are only at the beginning of your career as a SEAL. Ahead of you are more advanced training, placement, and deployments.

Advanced Training

The best people in their fields or discipline, whether they are astrophysicists or yogis, Olympic athletes or musicians, surgeons or Nobel-prize winning authors, have one thing in common— they never believe they have learned everything there is to know about what they do. They always consider themselves students, even when they have mastered their craft, *they are always learning.* SEALs remain students for their entire careers. Training never stops.

⌃ A squad of U.S. Navy SEALs pause for an after-action brief while participating in special operations urban combat training. The training exercise familiarizes special operators with urban environments and tactical maneuvering during day and night operations.

Your first duty station, as a new SEAL, will be either Virginia Beach, Virginia; Coronado, California; Pearl Harbor, Hawaii; or Panama City, Florida, to join the SDV Team. If you are an enlisted SEAL with a medical rating, meaning medical training or experience, you will report to Fort Bragg, North Carolina, for a six-month Advanced Medical Training course to become a Special Operations SEAL medic.

SEALs on the officer track must first attend a Junior Officer training course to learn operations planning and how to perform team briefings.

« **A member of SEAL Delivery Vehicle Team Two (SDVT-2) prepares to launch one of the team's SEAL Delivery Vehicles from a Dry Deck Shelter on the back of the Los Angeles-class attack submarine USS *Philadelphia* on a training exercise in the Atlantic Ocean.**

The SEAL Delivery Vehicle (SDV) is a manned submersible and a type of Swimmer Delivery Vehicle used to deliver United States Navy SEALs and their equipment for special operations missions. The SDV is used primarily for covert or clandestine missions to denied access areas, either held by hostile forces or where military activity would draw notice and objection.

Pre-deployment Work-Ups

Once you've reported to your assigned SEAL or SDV team, you will be joining a platoon for advanced training, also called "pre-deployment work-ups." The work-ups are in three phases lasting a total of roughly 18 months. During the first phase, you will receive specialty training, which can include foreign language study, SEAL tactical communications, military free-fall parachuting, explosive breacher, and/or sniper training.

During this time, the officers, senior enlisted, and other more experienced SEALs will be assessing you for special talents and skills.

★★★★★

"The purpose of fighting is to win. The sword is more important than the shield, and skill is more important than either. The final weapon is the brain, and all else is supplemental."

—John Steinbeck

As mentioned above, from their early beginnings, SEALs and their predecessors were tasked with some of the most challenging operations in the history of the U.S. military. The Navy has always required its operators to be in top physical condition and have an abundance of war fighting as well as other skills in order to meet the rigorous demands placed on them.

During the Korean War the early Underwater Demolition Teams (UDTs) conducted demolition raids on railroad tunnels and bridges along the Korean coast, supported mine-clearing operations, conducted beach and river reconnaissance, and infiltrated guerrillas behind enemy lines.

In 1962 the SEAL Teams were established and made up of men from the UDTs. The SEAL mission was to conduct counterguerilla warfare and clandestine operations in maritime and riverine environments. SEAL involvement in Vietnam began immediately after they were commissioned. In 1983, all UDTs were redesignated as SEAL Teams or Swimmer Delivery Vehicle Teams.

Naval Special Warfare (NSW) forces have since participated in Grenada 1983; Persian Gulf 1987–1990; Panama 1989–1990; Middle East/Persian Gulf 1990–1991; and have conducted countless missions in Afghanistan, Bosnia, Haiti, Iraq, Liberia, Somalia, and elsewhere.

During Operation Iraqi Freedom, Naval Special Warfare deployed the largest number of SEALs in its history. The SEALs were instrumental in numerous special reconnaissance and direct action missions including the securing of the southern oil infrastructures, clearing waterways, the capture of high value targets, raids on suspected chemical, biological, and radiological sites, and the first POW rescue since World War II.

The U.S. Navy SEALs are considered not only as the toughest but also the most versatile, fittest, and most lethal warriors on the planet. Navy SEALs not only have to be incredibly fit but must be able to operate under any condition and in every environment the planet has to offer including parachuting into enemy and hostile terrain; swimming in rough seas; diving in icy waters; and conducting maritime operations launching from submarines, small craft, ships, fixed wing aircraft, and helicopters. They operate in arctic, mountain, desert, and jungle terrain. Basically a Navy SEAL can operate efficiently anywhere in this world and can insert or extract from the area of operation (AO) in every possible means the military has to offer.

Any skill or expertise a SEAL can bring to the teams can be critical. Some SEALs come into the teams with extensive experience in rock or mountain climbing, emergency medicine, defensive driving, boating, salvage or deep sea diving, piloting small aircraft, mechanics, engineering, construction, computers, photography, weapons, languages, knowledge of foreign nations, and military history, to name just a few.

The expertise I brought to the teams was that I was an EMT with quite a bit of experience treating wounded personnel. I was also a mountaineer and climber which was of value during our mountaineering and rock climbing trips and oil rig, shipboard, and building climbing evolutions.

As a SEAL, I gained extensive experience in all of the SEAL skills we were required to maintain. I also received a great deal of advanced instruction and training in advanced and foreign weapons, sniping, Close Quarters Combat (CQC), Close Quarters Battle (CQB), mechanical and explosive breaching, advanced demolitions, surreptitious entry, advanced diving, jumpmaster, diving supervisor, High Threat Protective Security (PSD), advanced combat medicine, and language (German and Spanish).

I had a great deal of pride in having the skills to save lives. Throughout my career in the Navy I was often in the right place, at the right time, and possessed the right skills to save numerous lives. I really liked providing life-saving aid to those in need of critical emergency care.

A couple of days after the U.S. invaded Panama, we were ordered to report to Caribbean side of the canal where the U.S. Army had attacked a Panamanian fishing boat and killed everyone on board.

We arrived at around noon of a steamy hot day about twenty-four hours after the engagement.

An Explosive Ordnance Disposal (EOD) team had already searched the fishing boat for explosives but hadn't removed the remains of the boat captain and six PDF soldiers who were rotting in the sun. The smell, as you can imagine, was disgusting.

We had orders to clean up the 60-foot vessel and tow it into harbor. The captain of the boat had been shot in the head with a large caliber round. All that was left of his head was a small portion of his afro. Maggots were in the process of eating away what was left of his brains. Behind him against the cabin's wall was a large smear of blood.

I used this later when I trained guys in CQB. Since the captain had been standing next to a metal wall, it was easy to see that rounds had skipped off the flat surface and hit him. So stand away from hard surfaces during combat.

I said to the command Master Chief, the Warrant Officer and to the six SBU-26 guys who were with us, "Body remains throw them overboard. Marijuana, coke, or any drugs, just place it all in a pile. Anything that looks like it could be intel, put it over here."

We threw some of the marijuana in the water and watched the fish devour it.

While the SBU-26 guys were cleaning and disinfecting the boat with Lysol, the CWO and the Master Chief were doing their engineering and mechanical work to ensure the vessel was fit to transport. It was my job to inspect the hull, shaft, and screw to see if the underside of the boat had sustained any damage during the attack.

Typically in the Navy, we dove with a swim buddy. But I was the only one doing this inspection dive. So after I suited up, and started to make my way down the ladder I said to the Master Chief, "If you need me for anything just bang three times on the deck and I'll surface."

The Master Chief picked up something that looked like a small pipe and banged it on the deck of the boat.

"How's this?" he asked.

"Great."

My dive took about forty minutes. There wasn't any obvious damage to the hull, shaft, or screw, and though I hadn't heard any tapping, the Master Chief did tap a bit about the boat, taking soundings.

Soon after I surfaced, we set up a large GPL (general purpose large) tent about thirty meters from the ship. Opposite the tent, about fifty meters from the ship, was a large dumpster. Behind us stood the fire department which was adjacent to the Army Jungle Survival School.

We heard sporadic shooting and rocket explosions in the distance.

Once we got the tent up, I said, "Guys, throw those old bloodied uniforms in the dumpster, then take a break. You've been working hard."

A few minutes later three Army guys walked over to our PB (patrol boat)—Captain Mark Meisner, a major, and a sergeant.

Captain Meisner asked, "Do you mind if we take the stuff, the uniforms, you're throwing away because we're starting a war museum?"

The Master Chief said, "Sure. Help yourself."

Captain Meisner and the sergeant jumped in the dumpster and started handing stuff out to the major. Then the sergeant climbed out.

Sometime during the day that piece of "pipe" that the Master Chief had banged on the deck of the boat and then used to take soundings throughout the vessel ended up in the dumpster.

I was sitting in the tent at around 2145 hours when I heard a tremendous boom.

Captain Mark Meisner had leaned over to pick up the pipe, which apparently was really a live RPG round, and it had exploded.

All of us in the tent grabbed our weapons. I ran outside in the dark and saw trash was scattered everywhere.

The major and the sergeant were screaming, "HELP! Get help!"

We all had our first-line gear and second-line gear next to our cots—which included weapons—and I had my SF (Special Forces) medical kit, which was part of my second-line gear. I did what combat medics are trained to do in medical emergencies: first establish that the scene is safe.

The major and sergeant were screaming, but weren't hurt.

Captain Meisner lay on the ground outside the dumpster looking about as dead as a person could look. His left lower leg above the knee had been blown 100 meters away. His right hand and most of his left hand were gone. Both of his eyes were hanging out of the sockets by the optic nerves. And thousands of pieces of shrapnel had ripped into his face and body.

All of the combat medical training I'd received immediately kicked into place. I looked at his chest, and listened and felt for any signs of breath from his mouth and nose. I saw that his chest was barely rising, which meant that he had an airway and was still breathing. His upper left arm had a major arterial wound that was pumping out fresh bright red blood. I covered it with my hand and then quickly applied a blow-out patch.

His left femur was completely exposed. But with a traumatic amputation like that, the vessels constrict and seal up, so the bleeding was minimal.

The major knelt beside me. The sergeant, some Army Rangers, and all of the SBU (special boat unit) guys eagerly offered to assist me.

I opened my SF medical bag—which was a little bigger than a one-day back pack—and started issuing instructions.

I said, "You. Take this Kerlex and wrap up that leg."

"You, wrap that arm."

All the time I was talking to the Captain, asking, "Sir, we are going to take good care of you, you are in good hands, my buddy is going to place a bandage over your hand." And so on.

In a serious injury you need to calmly talk to the patient. Hearing is the last sense to go before death. Even when the patient doesn't respond, there's a chance that him hearing you will prevent him from going into a deeper stage of shock. In the unfortunate cases where the patient doesn't survive, at least the last thing they'll hear is a friendly soothing voice.

As in any trauma case, I constantly monitored the captain's AVPU scale—another great lesson from Goat Lab. AVPU (Alert, Voice, Pain,

Unresponsive) is a tool for assessing vital signs. If the person is Alert, he knows the approximate time and date, his name, and what caused the accident. In that case, I would make mental note or ask someone to record "Alert 1545."

The patient is one step lower on the scale if he or she can only respond to Voice. In other words, "Are you okay?"

"Yea."

Below that is Pain—response to painful stimuli. For example, you may hear the patient moan as you apply a splint or IV. I might also tweak the person's ear or rub their sternum with the knuckle of my middle finger to see if they respond.

The worse case, other than death, is when the victim is Unresponsive to voice or pain.

Initially, Captain Meisner was totally unresponsive. I tweaked his ear and did a sternum rub, but no response.

I monitored his airway constantly, because it can close at any time from bleeding, vomit, mucous, broken bones, teeth, or swelling. Just because the patient has an airway during your initial assessment doesn't mean that it won't close off during your treatment.

Captain Meisner had an airway, and we had stopped most of the bleeding. Now I needed to get fluids in as quickly as possible. Since he'd lost so much blood he needed a blood volume expander—Ringer's Lactate. But I had a hard time finding a good vein because all of his limbs were injured. I finally managed to get two large-bore IVs in his good arm, which pumped in 4000ccs of Ringer's Lactate.

Within twenty minutes, I'd gone through everything in my SF medical kit.

The captain's pulse was over 110 beats per minute. His heart was beating this fast because his brain was crying out for more oxygen.

If his pulse was 110 and his breathing was 20, and I check three minutes later and his pulse was 130 and his breathing was 30, then two minutes after that it was up to 160 and 40—that meant something was seriously wrong, and I needed to correct it quickly or the captain would die.

What had been killing Captain Meisner were all the leaks. Now that we had stopped them, his heart rate and breathing started to stabilize. Then he moaned and began moving around a little bit. He'd gone from

Unresponsive to Pain. A step in the right direction. His pulse was getting stronger and his breathing was becoming fuller, too.

Suddenly, he asked in a whisper, "Hey, what happened?"

I said, "Sir, what's your name?"

"Captain Mark Meisner."

"Do you know what just happened?"

"No, but I went to pick up a flashlight."

That told me that he saw the live rocket round in the dumpster and went to pick it up.

He asked, "How come I can't see?"

As a medic I couldn't say, *Because your eyes are hanging down your face.*

It was my job to keep him as comfortable as I could.

I said, "You're going to be fine, sir."

He asked, "How come I can't feel my hand?"

I said, "We'll look at that. We just called medivac."

Half an hour had passed since the explosion went off. It seemed that medivac was taking forever.

Captain Meisner asked, "How come I can't feel my leg?"

He got to the point where he was actually joking about his condition. He said, "At least if I still have my nuts, my wife will take me back."

The man's courage was amazing.

I'd moved to Captain Meisner's side and had left a guy I didn't know in charge of holding his head, which you always have to do in the case of a traumatic injury because of the possibility of damage to the patient's head, neck, or spine.

When I saw him starting to move the captain's head, I said emphatically, "Don't move his head!"

"It's okay," the man responded. "I'm an Army doctor."

"Well, if you're a doctor, how come I'm doing all this."

The Army doctor said, "His neck isn't broken. I can tell."

"You don't have x-ray fingers," I shot back. "So you can't tell. Hold his head still."

After forty-five minutes the medivac team arrived and took Captain Meisner away. He ended up being treated at the Army Burn Center in Texas.

Later he sent me a letter. He told me that he'd lost one eye, but had partial vision in the other. He was fitted with a prosthetic leg. Surgeons removed a toe off his foot and attached it to his hand so he has opposing pressure on one hand, which meant he could grasp objects and lift them up.

The experience reinforced an important lesson: Don't panic when you see someone who is grotesquely injured. Stay calm and proceed to treat them the way you've been trained to.

Afterwards the Army conducted an investigation. They found out that three LAAW rockets had been fired at the boat, and the EOD guys had only recovered two.

So the story came together. And I'm very happy to report that Captain Mark Meisner survived. As a matter of fact our families had Christmas dinner together last year. We have formed a lifelong bond.

Excerpted from *Inside SEAL Team SIX*

Your training will give you the necessary qualifications and designations that will allow you to perform as a vital member of your platoon. Together with fellow SEALs trained in their specialties, you will become an effective operational combat team.

Next comes unit level training, where you and your platoon gain expertise in core mission skills—land warfare, close quarters combat, small unit tactics, urban warfare, combat swimming, hostile maritime interdiction (operations that halt enemy forces or supplies, either through delay, disruption, or destruction on their way to battle and before they harm friendly forces), long-range target interdiction, special reconnaissance, and rotary and fixed wing (i.e., helicopters and planes) air operations.

⌃ **SEALs and divers from SEAL Delivery Vehicle Team Two swim back to the guided-missile submarine USS Michigan during an exercise for certification on SEAL delivery vehicle operations in the southern Pacific Ocean.**
Credit: U.S. Navy photo by Mass Communication Specialist 3rd Class Kristopher Kirsop.

During the third phase you will train at task group level with what is called supporting attachments or enablers—among them intelligence teams, medical teams, interpreters, linguists, and cryptologists (decoders), special boat teams, and explosive ordnance disposal teams. You will gain a good understanding of the importance of the connection between individual, unit, and task group training and how each one builds on another. What you learn individually is your contribution to your platoon, and your platoon becomes a vital part of larger operations.

Following the third phase of training, you, as part of a SEAL squadron, will conduct a final certification exercise or CertEx. The exercise consists of synchronizing troop operations under the umbrella of the Joint Special Operations Task Force. When the exercise is completed, your SEAL squadron is certified for deployment. Conducting real world missions in enemy territory will now take priority over training.

Individual Advance Training Courses and Schools (partial list)
- Sniper | Scout/Sniper
- Advanced Close Quarter Combat/Breacher (Barrier Penetration/ Methods of Entry)
- Surreptitious Entry (Mechanical and Electronic Bypass)
- NSWCFC (Naval Special Warfare Combat Fighting Course)
- Advanced Special Operations (MSO)
- Technical Surveillance Operations
- Advanced Driving Skills (Defensive, Rally, Protective Security)
- Climbing/Rope Skills
- Advanced Air Operations: Jumpmaster or Parachute Rigger
- Diving Supervisor or Diving Maintenance-Repair
- Range Safety Officer
- Advanced Demolition
- High Threat Protective Security (PSD)—(US/Foreign Heads of State or High Value Persons of Interest)
- Instructor School and Master Training Specialist
- Unmanned Aerial Vehicle Operator
- Language School
- Joint SOF and Service Professional Military Education (JPME)

Platoon Life

As explained at navyseals.com, the predeployment workup is "fast paced training, conducted throughout the United States by group level training detachment SEALs who have 'been there, done that,' and are experts in their specialty." On a workup you could find yourself somewhere practicing close quarters combat, only to speed off to another location to train in another skill—it could be anything.

Platoons are often out training for much of the time during workups, with short visits home and quick turn-arounds. During this time your platoon will constantly be tested. It is during this time, too, that a deeper, we're-all-in-this-together camaraderie between teammates— something highly prized among special operators—really takes form. These men must have your back, and you have theirs.

Operators from a West Coast-based Navy SEAL team exit an Army CH-47D Chinook during infiltration and exfiltration training as part of Northern Edge in Alaska. ⌃

Deployment

Deployment. You've finally made it. It may have taken more than a few years—often does. As a Navy SEAL you will work for United States Special Operations Command (USSOCOM) Army/Navy/Air Force Special Operations Force and, when necessary, for the Navy battle staffs on amphibious readiness groups or fleet ships, and aircraft carriers. All of this is to support the overall strategy of the highest military combatant commander—it goes from the Commander-in-Chief through the Secretary of Defense to the Combatant Commanders—think Generals Eisenhower and Schwarzkopf.

⌃ **U.S. Navy SEALs train with Special Boat Team 12 on the proper techniques of how to board gas-and-oil platforms during the SEALs gas-and-oil platform training cycle in Long Beach, California.**

Credit: U.S. Navy photo by Mass Communication Specialist 3rd Class Adam Henderson.

⩘ **U.S. Navy SEALs fast-rope onto the fantail of the guided missile destroyer USS *Oscar Austin*.**

Credit: U.S. Navy photo by Photographer's Mate 1st Class Michael W. Pendergrass.

⩘ **A Navy SEAL fires an MK-11 sniper rifle from an MH-60S Sea Hawk helicopter assigned to Helicopter Sea Combat Squadron 9, deployed aboard the aircraft carrier USS *George H. W. Bus*h during a training flight over the Arabian Sea.**

⌃ Petty Officer Second Class (SEAL) Michael A. Monsoor patrols the streets of Iraq while deployed in 2006.

For his actions in Ar Ramadi, Iraq, Monsoor received the Medal of Honor posthumously in a ceremony at the White House in April 2008. As stated on his official Navy biography, on September 29, 2006, "Monsoor was part of a sniper overwatch security position with three other SEALs and eight Iraqi Army (IA) soldiers. An insurgent closed in and threw a fragmentation grenade into the overwatch position. The grenade hit Monsoor in the chest before falling to the ground. Positioned next to the single exit, Monsoor was the only one who could have escaped harm. Instead, he dropped onto the grenade to shield the others from the blast. Monsoor died approximately 30 minutes later from wounds sustained from the blast. Because of Petty Officer Monsoor's actions, he saved the lives of his 3 teammates and the IA soldiers."

He had also received the Silver Star for earlier actions during the same deployment, when he exposed himself to heavy enemy fire to rescue and treat an injured teammate.

⤝ Navy SEALs and special warfare combatant-craft crewmen recover parachutes and personnel during a maritime craft aerial deployment system exercise near Fort Walton Beach, Florida.

Credit: U.S. Navy photo by Mass Communication Specialist 2nd Class Joseph M. Clark/Released

« **Members of SEAL Delivery Vehicle Team Two inside a flooded Dry Deck Shelter mounted on the back of the Los Angeles-class attack submarine USS** *Philadelphia.*

Credit: U.S. Navy photo by Chief Photographer's Mate Andrew McKaskle.

≽ U.S. and Polish special operations forces practice boarding skills during an exercise to further develop interoperability skills as part of the USEUCOM active security strategy near Gdansk, Poland.

≽ U.S. Navy SEALs conduct a fast-rope insertion demonstration from an MH-60S Sea Hawk helicopter into the Chesapeake Bay during a capabilities exercise at Naval Amphibious Base, Little Creek, Virginia.

Credit: U.S. Navy photo by Mass Communication Specialist 2nd Class Joshua T. Rodriguez.

☆ **Navy SEALs conduct close quarters combat training in a simulated home at U.S. Training Center Moyock in North Carolina.**

U.S. Navy photo by Mass Communication Specialist 2nd Class Eddie Harrison.

☆ **A Navy SEAL climbs into the turret gunner position during a mobility training exercise through a simulated city.**

Credit: U.S. Navy photo by Mass Communication Specialist 2nd Class Eddie Harrison.

⚹ A U.S. Navy SEAL team helps secure the airfield as Air Force One lands at Al Asad Air Base, Iraq.

⚹ Chief Special Warfare Operator David Goggins climbs the cargo net obstacle with a Make-A-Wish Foundation recipient in San Diego. The young man had undergone a heart transplant and realized his dream of being a Navy SEAL as he toured Naval Special Warfare facilities with Goggins, who also endured two heart surgeries during his career.

Credit: U.S. Navy photo by Mass Communication Specialist 2nd Class Dominique Lasco.

⌃ U.S. Navy SEALs lead exercises at Cooper High School in Minneapolis, Minnesota. The SEALs also held a "Mental Toughness" presentation to promote Naval Special Warfare awareness.

Credit: U.S. Navy photo by Mass Communication Specialist 2nd Class William S. Parker.

⌃ A U.S. Navy SEAL sniper shows Naval Special Warfare ombudsmen and family support advocates the weapons SEALs use while deployed during the fifth annual Ombudsman Conference in San Diego.

Credit: U.S. Navy photo by Mass Communication Specialist 2nd Class Shauntae Hinkle-Lymas.

☆ **A SEAL delivery vehicle (SDV) team performs a fast-roping exercise from a MH-60S Seahawk helicopter to the topside of Los Angeles-class fast attack submarine USS *Toledo*. SDV teams use "wet" submersible vehicles to conduct 100 percent long-range submerged missions, or to secretly deliver SEALs and other agents onto enemy territory from a submarine or other vessel at sea.**

Credit: U.S. Navy photo by Journalist 3rd Class Davis J. Anderson.

You will be deployed to one of four geographic areas of operation (AO)—CENTCOM (Middle East); SOUTHCOM (Central and South America); PACOM (Pacific and Korea); EUCOM (Europe)—to carry out a mission.

It may involve insertion into a combat objective by parachute, helicopter, submarine, high-speed boat, combat swimming, or on foot. You will likely operate a wide variety of high-tech, specialized equipment and weapons.

You will very likely spend at least one deployment in the Middle East. For operational security reasons, we cannot furnish specific details on deployment schedules, rotations, or other sensitive or classified information on *any* mission. But plenty of general information about core missions can be found in chapter 1.

Panama

"General Noriega's reckless threats and attacks upon Americans in Panama created an imminent danger to the 35,000 American citizens in Panama. As President, I have no higher obligation than to safeguard the lives of American citizens."
—President George H. W. Bush announcing Operation Just Cause, December 20, 1989

Panama's narrow neck of tropical land is intersected by one of the most important strategic waterways in the world—the Panama Canal. Panama was run by a short, pugnacious military dictator named General Manuel Noriega.

Noriega, who had operated as a CIA asset, bought the loyalty of Panamanian Defense Force by growing a criminal economy that operated as a parallel source of income for the military and their allies, providing revenues from drugs and money laundering.

Under Noriega's rule, Panama became the major transshipment site for illegal drugs from South America bound for the United States. While elements in the PDF prospered, Noriega's regime grew increasingly repressive with hundreds of political opponents of his regime tortured and killed. And hundreds more forced into exile.

Political demonstrations against the regime were met with violence. A vocal critic named Hugo Spadafora was pulled off a bus by Noriega's men at the Costa Rican border. A subsequent conversation between Noriega and the local PDF commander, Luis Cordoba, was captured on wiretap.

Cordoba: "We have the rabid dog."

Noriega: "And what does one do with a dog that has rabies?"

Several days later Spadafora's badly tortured, decapitated body was found wrapped in a U.S. Postal Service mail bag.

On February 5, 1988, General Noriega was accused of drug trafficking by federal juries in Tampa and Miami.

President George H. W. Bush protested loudly, demanding that Noriega end political repression and drug trafficking, and expressing concern about the secure functioning of the Panama Canal, which was vital to U.S. shipping and regional security.

Sometime in early 1989, the CIA, Naval Special Warfare, and other government and military units were starting to collect intel in Panama for a possible op to arrest Noriega and oust him from power.

I arrived in Panama in November 1989 shortly after a coup led by Panamanian Major Moises. The moment I landed, I felt the tension in the air. I heard the stories about Noriega's out-of-control cocaine parties, and about him throwing people out of planes.

I, and a handful of other SEALs from ST-6, was stationed at Rodman Naval Station, which bordered the west side of Panama Canal. We were there to assist the Special Boat Unit 26 (SBU-26) with coastal and riverine operations. I also had orders to serve as the Training Officer, and to establish and run the Navy Special Warfare Jungle Training.

On the night of December 16, 1989, I was driven in an armored vehicle to the airport to meet my girlfriend Shannon, who was flying in to spend Christmas with me. Accompanying me was a young SEAL lieutenant also assigned to Special Boat Unit 26 named Adam Curtis, whose wife Bonnie was coming in on the same flight—very rare occasions for any of us, especially in a hot zone.

The tension at the airport was palpable. You could see the anxiety about a possible U.S. military action in peoples' eyes.

After the plane carrying the women landed, Adam informed me that he was taking his wife out to dinner.

I said, "Be careful, it's getting bad out there. Shannon and I are going back to the base."

Adam Curtis and Bonnie had finished dining at a local restaurant and were on the way back to his barracks when they were stopped at a PDF checkpoint. They were questioned and their car was searched.

Adam later remembered that "While we were there, another group of Americans came to the road block, three Army guys and a Marine—all officers. They felt threatened, they gunned it through the roadblock, and five PDF soldiers turned and fired at the car. The officer in the back, an Army lieutenant named (Robert) Paz, was killed."

Adam Curtis and his wife Bonnie were pulled out of their car and taken to a detention center, where they were interrogated and tortured. While PDF goons hammered Adam's feet in one room, they fondled and sexually harassed his wife in another.

The next morning at muster, Adam wasn't there.

The Captain turned to me and asked, "Where's Lieutenant Curtis?"

"I don't know," I answered. "Last time I saw him was last night at the airport."

According to various sources, President George H. W. Bush made the final decision to invade Panama after hearing about the murder of Lt. Robert Paz and the detention and torture of the Curtises.

By Sunday the 20th, elements of Navy SEAL Teams 2, 4, and 6 had infiltrated the country. At around midnight, elements of SEAL Teams 2 and 4 were on the move.

SEAL Cmdr. Norman Carley, task unit commander for SEAL Team 2 and former executive officer to Richard Marcinko at ST-6, was with his men aboard combat rubber raiding craft (CRRC) in a mangrove, waiting to launch four SEAL swimmers to attach a limpid mine to the *Presidente Porras* patrol boat. "The commander of the whole operation, Gen. (Carl) Steiner, thought that the operation had been compromised, and he moved up the time to execute the operation by a half hour," Carley recalled. "But the clocks and safety and arming devices on the explosives were already set."

"As they were doing so (attaching explosives to the patrol boat), a firefight was going on and grenades were falling into the water," said Carley. "They (SEALs) thought that they were detected. But Randy B. had just finished attaching their explosive devices before swimming away." As the clock struck 0100, a large blast from the SEAL's explosives shook the walls of buildings across Panama, sending PDF soldiers scrambling for an imminent battle. That part of the mission was a big success—the first time SEALs successfully executed a limpid attack on an enemy ship of battle.

I was in base housing at Fort Amador, which was about four miles from Rodman, when Shannon and I heard an AC 130 gunship firing rounds at a target in Panama City. I telephoned work and they said, "Get in here immediately!"

I grabbed my gear and weapon and started running as fast as I could toward Rodman Naval Station. I was so excited!

A U.S. Army jeep speeding down the road saw me running and stopped. An MP asked, "Where the hell are you going?"

"My name is Chief Mann and I am on my way to Rodman. I need a ride."

They rushed me to Rodman. Minutes later I was on a river patrol boat (RPB) with six guys from SBU-26. SBU-26 was commanded by a Navy Lieutenant Commander (O-4) Mike F., a tough Vietnam-era SEAL with more riverine experience than anybody else on the teams. It was made up of a headquarters element and ten Patrol Boat Light detachments. Each detachment consisted of two boats with crews.

Our first frag order of the night was to confirm the reported sniper fire that was coming from the Bridge of the Americas.

We fired up the PRB (Patrol River Boat), basically a beefed up Boston Whaler armed with MK-19s and twin 50s, and approached the bridge. As we emerged from the shadows we started taking fire. The sniper had the advantage of concealment and elevation. With rounds ripping into the water around us, we trained our twin mounted .50 caliber machine guns on the snipers. They ducked behind some metal beams and fled.

At the same time a few miles away, SEALs from ST-4 were coming ashore in small inflatable boats near the Punta Paitilla Airfield. Their mission was to seize the small civilian airfield and disable Noriega's Learjet so he couldn't escape. But the element of surprise had been eliminated because of the explosion on Noriega's yacht. Also, the runway was well lit by landing lights and the AC-130 Spectre gunship that was assigned to provide air cover was unable to launch.

Those weren't the only problems the SEALs encountered. They had been told that the airstrip wasn't guarded. But the intel was wrong.

Dennis H., a lieutenant at the time, was the platoon officer in charge. "As we advanced, I heard yelling," Hansen remembered. "The plan was to tell the Panamanian security guards to go away. This seemed to work well until we got to Noriega's plane hangar. There, a gunfight broke out after a brief exchange of words. The platoon adjacent to mine was directly in front of the hangar. They were to disable the plane. About half of the platoon was wounded. I sent my assistant officer in charge (AOIC) and his squad to support the platoon that was in contact. They took effective fire also, killing my AOIC and wounding a couple of other men."

Four SEALs died in the firefight: Lt. John Connors, my good friend, Chief Engineman Donald McFaul, Boatswain's Mate 1st Class Chris Tilghman, and Torpedoman's Mate 2nd Class Isaac Rodriguez III.

Another very good friend of mine named Carlos Moleda was shot in the chest and leg. Another teammate, Mike P., thought Carlos was dead and used his body as a shield as he returned fire.

I was with him when he apologized. He said, "Sorry, Carlos, but I thought you were dead and I couldn't get low enough, and needed anything in front of me to block me from the incoming fire."

Fortunately, Carlos survived. Even though he never recovered use of his body from his sternum down, Carlos went on to compete in several Ironman and ultra distance athletic events in his wheelchair and has won many!

Our next assignment was to capture Noriega's yacht, which was docked on the south side of the canal. The SBU-26 executive officer, LCDR (0-4) Johnny K., who could out-PT, out-swim, and out-run most SEALs who were twenty years younger, received the order and wanted to accompany us.

The general in charge ordered him to stay.

Meanwhile, our Patrol Boat (PB) had almost finished backing away from the pier and was turning right. Johnny hung up the phone, ran down the pier, jumped in the water in full uniform, web gear, and weapon (M-16) and swam to the boat. Johnny wasn't going to let anybody tell him that he couldn't go into battle with his men.

The crew helped him on board the vessel and he went on the op.

Noriega's yacht was at least forty feet long with quarters for at least eight, ocean fishing rods and reels, and a wine cellar. We pulled up within fifty meters and observed the vessel with our .50 cals locked, loaded, and aimed at the craft. We had no idea how many people were on board, or if the yacht had been booby-trapped with explosives.

Our Spanish speaker got on the horn and announced that those onboard had thirty seconds to surrender before we blow their boat out of the water. About ten seconds later a hatch to the lower deck opened, and a guy stuck his arm out and waved a white t-shirt.

I was the first to board with my M-16. Three other SBU-26 guys followed behind me. I saw a group of eight Panamanian SEALs hiding in the cabin and motioned for them to drop their weapons and come out.

Through our Spanish speaker, I ordered them to strip down. Since there were so many of them and the deck was small, I directed them to stack on top of one another, head to toe.

When one of the Panamanians refused, Johnny K. yelled to our Spanish speaker, "Tell him to strip down now, or we'll shoot him!" The Panamanian complied. We tie-tied them and hauled them back to Rodman where we made a makeshift prisoner compound out of barbed wire. We'd cleared the yacht, and before the night was over we captured around 200 PDF enemy soldiers.

At around midnight (eleven hours after the launch of the invasion) we approached within 500 meters of another enemy vessel with our weapons ready. When we moved within 200 meters, the interpreter yelled in Spanish, "Come out, or we're going to blow you out of the water!"

The captain said, "Let's move a little closer." We pulled to within 50 meters going bow to broadside. I was thinking, *This is insane! We're getting too close. A firefight between the two crews is going to be brutal.*

We were so close we could almost smell them. Suddenly a hand holding a little white handkerchief emerged from one of the cabin windows, and we gave a collective sigh of relief.

We captured another dozen armed PDF soldiers.

Little by little we gained control of the Panama Canal—blocking boats from entering, and stopping boats on the canal and searching for arms and enemy personnel. Meanwhile, U.S. Army and Marine battalions supported by airpower attacked the PDF's central headquarters (*La Comandancia*) in downtown Panama City and seized Fort Amador from the PDF in a nighttime air assault.

At some point during the night we exchanged fire with a half dozen PDF soldiers along the shore and tore them up pretty bad. We put the bodies in body bags, but when we returned to Rodman, nobody knew where to put them. So we decided to stash the bodies in the meat locker at the Rodman Enlisted Club with the steaks, hamburgers, fruits, and vegetables.

I found this somewhat disturbing because I ate lunch there almost every day.

When the U.S. military documentation people arrived a couple of days later, they asked me to open two of the body bags and turn the bodies over so they could photograph the faces. These particular bodies had been riddled with bullets. When I pulled one of them from the side to flip it over, the top half pulled away and separated from the bottom. It left me with a very unpleasant image that I'll never forget.

Excerpted from *Inside SEAL Team Six*

Standing Down

After platoons return from deployment they go through the "stand down" phase, turning in weapons, equipment, gear, and supplies to be inventoried and repaired for reissue during the next deployment. Workdays are short, and for a reason. Stand down, which lasts about a month, is a time for SEALs to spend with family and enjoy some down time.

More about Missions and Operations

If you picked up this book we suspect that even before you cracked it open you have already heard or read about enough special forces missions and operations to have sparked a deep interest in the idea of becoming a SEAL.

Yes, a few operations have been headlined in the media—the rescue of the mission workers in Somalia; the retrieval of the kidnapped ship's captain from pirates; the execution of Osama bin Laden. The men who carried out those missions deserve all the credit in the world. Because within all the seeming glamour and excitement surrounding these missions are men who not only went through everything outlined in the earlier chapters of this book and more, but did it under grindingly difficult conditions, under the radar, and without any other objective than to become an outstanding special forces operator.

But what really sets them apart is that they are willing to undertake *any* mission or operation handed to them, not just the "glamorous" ones. The SEAL ethos compels them to. As a member of the SEAL teams you must *want* to carry out your missions and carry them out as flawlessly as possible.

Your missions as a SEAL will also include noncombat operations: community outreach and copartnering with local fire and police departments as well as participation in local school sporting and charitable events.

Pay and Benefits

A career as a Navy SEAL provides intangible and tangible benefits. SEALs work closely with their teammates to innovatively accomplish missions important to national security. They constantly learn and push their physical and mental limits and live a very unconventional lifestyle.

Navy SEALs operate at the forefront of America's national security efforts by conducting direct action missions against high value terrorist cells. SEALs have provided security for key officials, taught foreign military and special operations personnel how to combat terrorism in their own countries, and prevented environmental sabotage. Wherever they are assigned, Navy SEALs make a difference by serving their country and protecting both U.S. and global citizens.

Every teammate on a SEAL Team brings his unique identity, personality, intelligence and talents to contribute to the SEALs' collective abilities to rapidly adapt and innovate on the battlefield. Each man who wears the SEAL Trident has endured the same extreme mental and physical challenges, making the SEAL Teams the ultimate brotherhood. Teammates do not let each other down, on or off the battlefield. The relationships and friendships formed within "The Teams" last a lifetime.

Naval Special Warfare is a lifestyle rather than just a job or career. "The office" for a Navy SEAL transcends not only the elements of the sea, air and land, but also international boundaries, the extremes of geography and the full spectrum of conflict. There is no "typical day at the office" for a Navy SEAL.

All active duty military personnel receive:

- a salary
- medical and life insurance
- 30 days annual leave
- education funding
- medical and dental benefits
- retirement after 20 years
- tax-free pay in combat zones
- tax-free allowances for housing and food
- access to military facilities
- travel and supply discounts

> In addition, SEALs receive:
>
> - starting pay of up to $60,000
> - qualification and re-enlistment bonuses
> - extra pay for diving, parachuting, and demolitions
>
> **From *www.sealswcc.com,* an official
> Naval Special Warfare site.**

Women in the Navy SEALs

Although the military ban on women in combat has been lifted, as of the publication of this book, there are no women in the Navy SEALs. Still, in June 2013 the Pentagon announced plans that would allow women to train for the Army Rangers by mid-2015 and the Navy SEALs in 2016.

"It's time to do this," said Adm. William McRaven, head of Special Operations Forces at the Special Operations/Low Intensity Conflict conference in Washington, D.C. "We've had women supporting direct Special Operations for quite some time," he added. "The one thing we want to make sure [we do is] maintain our standards."

The concern about this is a serious one given the rigorous physical demands that Special Operations Forces require of all its members. Lowering those standards is not an option. Col. Ingrid Gjerde, an officer in the Norwegian infantry for 25 years and commander of Norwegian forces in Afghanistan in 2012 said, in an article in the *Christian Science Monitor* in January 2013, that female troops in Norway have fought to keep the standards at the level maintained for men.

In January 2013 Joint Chiefs of Staff Chairman Gen. Martin Dempsey said that banning women from certain combat jobs has added to a sexually hostile environment for females in the military:

"I believe it's because we've had separate classes of military personnel, at some level. Now, you know, it's far more complicated than that, but when you have one part of the population that is designated as warriors and another part that's designated as something else, I think that disparity begins to establish a psychology that in some cases led to that environment. I have to believe, the more we can treat people equally, the more likely they are to treat each other equally."

AFTERWORD

This handbook is meant only as a tool and resource on your way to becoming a member of the Navy SEAL teams. It will be up to you and only you to make that happen. We certainly hope you found the information here helpful in deciding if the SEAL teams are for you.

Hoo Ya!

—Don D. Mann
CWO3/Retired
SEAL Team ONE
SEAL team TWO
SEAL Team SIX

ASVAB SAMPLE QUESTIONS
(FROM TODAYSMILITARY.COM)

General Science

General Science tests the ability to answer questions on a variety of science topics drawn from courses taught in most high schools. The life science items cover botany, zoology, anatomy, and physiology and ecology. The earth and space science items are based on astronomy, geology, meteorology, and oceanography. The physical science items measure force and motion mechanics, energy, fluids, atomic structure, and chemistry.

Sample test question:

1. An eclipse of the sun throws the shadow of the

A. moon on the sun

B. moon on the earth

C. earth on the sun

D. earth on the moon

Arithmetic Reasoning

Arithmetic Reasoning tests the ability to solve basic arithmetic problems encountered in everyday life. One-step and multistep word problems require addition, subtraction, multiplication, division and choosing the correct order of operations when more than one step is necessary. The items include operations with whole numbers, operations with rational numbers, ratio and proportion, interest, percentage and measurement. Arithmetic Reasoning is one factor that helps characterize mathematics comprehension, and it also assesses logical thinking.

Sample test question:

2. How many 36-passenger buses will it take to carry 144 people?

A. 3

B. 4

C. 5

D. 6

Word Knowledge

Word Knowledge tests the ability to understand the meaning of words through synonyms—words having the same or nearly the same meaning as other words. The test is a measure of one component of reading comprehension since vocabulary is one of many factors that characterize reading comprehension.

Sample test question:

3. The wind is variable today.

A. mild

B. steady

C. shifting

D. chilling

Paragraph Comprehension

Paragraph Comprehension tests the ability to obtain information from written material. Students read different types of passages of varying lengths and respond to questions based on information presented in each passage. Concepts include identifying stated and reworded facts, determining a sequence of events, drawing conclusions, identifying main ideas, determining the author's purpose and tone, and identifying style and technique.

Sample question:

4. Twenty-five percent of all household burglaries can be attributed to unlocked windows or doors. Crime is the result of opportunity plus desire. To prevent crime, it is each individual's responsibility to:

A. provide the desire

B. provide the opportunity
C. prevent the desire
D. prevent the opportunity

Mathematics Knowledge

Mathematics Knowledge tests the ability to solve problems by applying knowledge of mathematical concepts and applications. The problems focus on concepts and algorithms, and involve number theory, numeration, algebraic operations and equations, geometry, measurement and probability. Mathematics knowledge is one factor that characterizes mathematics comprehension; it also assesses logical thinking.

Sample test question:

5. If X + 6 = 7, then X is equal to
A. -1
B. 0
C. 1
D. 7/6

Electronics Information

Electronics Information tests understanding of electrical current, circuits, devices, and systems. Electronics Information topics include electrical circuits, electrical and electronic systems, electrical currents, electrical tools, symbols, devices, and materials.

Sample test question:

6. Which of the following has the least resistance?

A. wood

B. iron

C. rubber

D. silver

Auto and Shop Information

Auto and Shop Information tests aptitude for automotive maintenance and repair, and wood and metal shop practices. The test covers several areas commonly included in most high school auto and shop courses, such as automotive components, automotive systems, automotive tools, troubleshooting and repair, shop tools, building materials, and building and construction procedures.

Sample test question:

7. A car uses too much oil when which of the following parts are worn?

A. pistons

B. piston rings

C. main bearings

D. connecting rods

Mechanical Comprehension

Mechanical Comprehension tests understanding of principles of mechanical devices, structural support and properties of materials. Mechanical comprehension topics include simple machines, compound machines, mechanical motion, and fluid dynamics.

Sample test question:

8. Which post holds up a greater part of the load?

A. post a

B. post b

C. both equal

D. not clear

Sample Question Answers:

1. General Science: B
2. Arithmetic Reasoning: B
3. Word Knowledge: C
4. Paragraph Comprehension: D
5. Mathematics Knowledge: C
6. Electronics Information: D
7. Auto and Shop Information: B
8. Mechanical Comprehension: A

NUTRITION AND TRAINING GUIDES

Nutrition is one of the four pillars of an effective fitness program; the others are exercise, recovery, and consistency. All four are necessary to becoming a SEAL, but nutrition is at least one-half of the equation to success. Exactly what type of diet will lead to your personal success has everything to do with your unique physiology. Every body responds differently to diet, training, and supplements.

We hope you find these guides—Special Operations Nutrition, Naval Special Warfare Physical Training, and Navy Special Warfare Injury Prevention—helpful in your effort to become a Navy SEAL.

From The Special Operations Forces Nutrition Guide

1 The Warrior Athlete

Special Operations Forces (SOF) are "Warrior Athletes," the ultimate athlete. The physical and mental demands imposed by SOF training and missions require appropriate nutritional habits and interventions so that, under the most rigorous conditions, performance is optimized, and health is preserved.

This chapter serves as an introduction to the specialized needs of SOF and the information to be presented in *The Special Operations Forces Nutrition Guide*.

2 Balancing the Energy Tank

- Balancing energy intake and expenditure can be difficult when activity levels are very high and also when activity levels are very low, such as during isolation.
- Typically, body weight remains constant when energy intake equals expenditure.
- To lose or gain one pound of weight, 3,500 calories must be expended or consumed.
- Calculating Resting Energy Expenditure (REE) and the intensity of daily activities gives an accurate estimate of how much energy an operator might expend in one day.
- The Body Mass Index (BMI) is a clinical tool for assessing body fat composition and

classifies individuals into underweight, normal, overweight, and obese categories.

3 Fueling the Human Weapon

- Carbohydrates (CHO) are the vital fuel for endurance and resistance activities, competitive athletic events, mental agility, and healthy living.
- Fats, the primary form of stored energy, are essential, but should be eaten in moderation.
- Proteins are essential for building and repairing body tissues; however, excess protein is converted to fat.
- Restore fluid balance by taking in enough liquids to replenish weight (pounds) lost plus an additional 25 percent.
- Performance decrements begin when only 2 percent of body weight has been lost.

4 High Performance Catalysts

- Vitamin and mineral needs can be met by eating a variety of foods.
- Vitamin-mineral supplements do not provide energy.
- Vitamin-mineral supplementation is warranted only when energy balance is not met through the diet.
- Mega-dosing on vitamins and minerals can be detrimental to health and performance.
- Foods naturally high in antioxidants (fresh and colorful foods) should be eaten daily.

5 Nutrient Timing and Training

- The timing of nutrient delivery is critical to sustaining performance.
- The **Refueling Interval (RFI)** is the 45 minutes after finishing a workout.
- Eating during the RFI will accelerate recovery and restore energy for the next day's workout.
- A daily diet that is balanced and nutrient-dense will ensure better performance and optimal recovery.
- CHO foods and beverages that have a moderate to high glycemic index, such as sport drinks, raisins, honey, bananas or potatoes are ideal recovery foods.
- Adding protein to the recovery meal will help stimulate protein synthesis to assist in rebuilding muscle (anabolism).
- For exercise longer than 90 minutes, consume 50 grams of CHO and 12 grams of protein as food or drink immediately during the RFI and 50 grams of CHO every 2 hours for 6 hours.
- Adequate fluids must be ingested after a mission.
- Fluid replacement beverages should contain sodium and potassium.
- Sports bars, gels and drinks are lightweight, portable and easy to eat during SOF operations.

6 Optimal Choices for Home Chow

- Foods eaten at home can impact mission performance.
- Smart shopping is the first step toward healthy meal preparation.
- Most recipes can be modified to improve nutrient composition.
- Use nutrition labels as a guide for making smart food choices.
- Every meal is important for overall health and performance.
- Aim for as many servings of fruits and vegetables as possible.

7 Optimal Choices for Eating Out

- Not all restaurants are equal. Choose wisely.
- Eating out can be healthy if careful meal selections are made.
- Selecting fruits and vegetables as a part of the meal adds vitamins, minerals, and fiber, and helps reduce fat and calories.
- Fast food restaurants have healthy alternatives to the high-fat burger and fries. Make sensible food choices.

8 Healthy Snacking

- Snacking, or "eating between regular meals," is important to help maximize performance and maintain mental and physical acumen.
- Healthy snacks can help increase energy and alertness without promoting weight gain.

- Keep nutrient-dense snacks at home, work, or "on the go."
- Snacks for night operations should include foods low in carbohydrate and high in protein.
- Snacks high in water, such as fruit, are great for warm weather operations.
- Snacks high in carbohydrate are good to consume when exercising in the cold.
- Avoid high-fat snacks during special operations.

9 Secrets to Keeping Lean as a Fighting Machine

- Consumption of carbohydrate (CHO) in defined amounts is the most important fuel strategy for all forms of exercise.
- Depletion of glycogen stores will result in poor performance in the weight room and endurance training sessions, such as a pack run.
- Improper nutrient intake and low muscle glycogen stores may increase the risk of musculoskeletal injuries.
- CHO ingestion improves the use of amino acids when they are ingested together.
- Drinking too much plain water can pose performance pitfalls during prolonged missions/exercise sessions that involve constant movement.

- Individual food preferences should be determined to avoid gastrointestinal distress during training and operations.

10 Bulking Up

- Proper and consistent strength training, adequate rest, and a balanced diet will provide the lasting **edge** when it comes to building strength and muscle mass.
- Eating a wide variety of foods and matching energy intake with energy output will provide optimal nutrition for building muscle.
- All operators require no more than 1 gram of protein per pound of body weight per day.
- Adequate amounts of fluids are vital to muscle metabolism and contractility.
- Spend money on "real" foods, **not** supplements and protein powders.

11 Looking for the Edge—Dietary Supplements

- SOCOM has a **no** dietary supplement (DS) policy—check with medical.
- DS sold on military installations are not always safe, effective, or legal.
- Manufacturers of DS are not required to conduct research on safety or effectiveness. The Food and Drug Administration must prove a product is unsafe before it can be taken off the market.

- If you use DS, select high quality products with USP (United States Pharmacopeia) certification labels. The label ensures consumers that the product has been tested and verified in terms of its ingredients and manufacturing process.
- Combining and stacking of DS increases the potential for undesired and unsafe side effects.
- Energy drinks are not regulated and the long-term effects of their combined ingredients are unknown.

12 Enemy Agents

- All tobacco products, including smokeless tobacco, are addictive and can cause cardiovascular damage and various forms of cancer.
- Alcohol, in excess, can lead to dehydration and compromise performance. Do not mix drugs and alcohol: beware of alcohol-drug interactions.
- Over-the-counter drugs, such as antihistamines, non-steroidal anti-inflammatory drugs, and aspirin should be used in moderation and under a physician's care if being used for long-term therapy.
- NSAIDs should not be used during deployments because they make bleeding difficult to control.
- Steroids and steroid alternatives are illegal and unsafe; they can seriously harm the body and negatively affect performance.

13 Combat Rations

- Combat rations are specially designed to supply adequate energy and nutrients for particular types of missions.
- Environmental and operational conditions dictate changes in combat rations to meet nutritional needs.
- Rations provide different amounts of energy to meet the needs of various operational conditions.
- Some rations have been designed to meet strict religious diets.
- Commercial products are available to supplement military rations and/or allow for greater diversity and choice for eating when deployed.

14 Eating Globally

- Be aware of cultural differences including types of food and proper eating utensils.
- Avoid foodborne illnesses by taking extra precautions: stay away from typical foods associated with foodborne illnesses.
- Make wise food and beverage selections when eating on the economy.
- Drinking contaminated water may severely affect your health: purify your water!
- Carry Pepto-Bismol and seek medical treatment for symptoms from contaminated foods or beverages.

15 Mission Nutrition for Combat Effectiveness

- Inadequate energy intake and/or dehydration can result in fatigue and impaired performance during combat.
- Improper eating and sleeping due to all night and high op-tempo missions can be detrimental to overall health.
- Eating before night operations should be planned accordingly to prevent fatigue.
- Various environmental exposures (i.e., heat, cold, and altitude) can alter combat effectiveness if nutritional needs and hydration are not met appropriately.
- Energy and fluid requirements are typically higher than normal during combat and combat-simulated scenarios.

16 Returning to Home Base

- Rest and rejuvenation should be emphasized upon return from deployment to re-optimize mental and physical performance.
- A good night of sleep in a comfortable bed and dark room is essential for recovering from deployments.
- A balanced diet high in complex carbohydrates, such as vegetables, fruit and whole grains, can enhance stress resistance.
- Good nutrition and regular exercise are excellent antidotes to stress.

- Avoid binge eating and drinking upon returning from deployments. Excess food and alcohol intakes can lead to unwanted weight gain and is detrimental to overall health.

17 The High Mileage SOF Warrior

- Try to maintain weight to minimize weight cycling– multiple episodes of weight loss.
- Pain from arthritis can be reduced by choosing healthy foods and foods high in anti-inflammatory compounds.
- NSAIDs should be used on a very limited basis.
- Foods, not supplements, should be the primary source of nutrients. Food is the best and cheapest way to take in essential nutrients.
- The risks of developing hypertension, coronary heart disease, diabetes, and cancer increase with age. Eating the right type of foods can limit risk factors associated with these chronic diseases.

18 Sustaining Health for the Long-Term Warrior

- Eating a variety of foods is one key to healthy living.
- A Mediterranean Diet has been shown to confer a long, healthy life.
- Healthy bones require adequate calcium intake and regular physical activity.

- Eating many different colorful real foods, which contain important protective compounds—phytonutrients, promote life-long health.
- At least 3-5 servings of colorful vegetables, 2 or more servings of fruit, and 6 or more servings of whole grain products, should be consumed per day, whenever possible.
- Products containing probiotics (yogurt, keifer, sauerkraut) may be helpful for maintaining a healthy digestive tract.
- Alkaline-forming, rather than acid-forming, foods are important during periods of high stress.

Find the complete guide at www.sealswcc.com/PDF/special-operations-nutrition-guide.pdf

Naval Special Warfare Physical Training Guide

The Naval Special Warfare Physical Training Guide is designed to assist anyone who wants to improve his fitness in order to take and pass the Physical Screening Test (PST) and succeed at Basic Underwater Demolition/SEAL (BUD/S).

This guide provides information about the type of training required to properly prepare for the rigors of BUD/S, and it offers a tailorable 26-week training plan that should help a person with average fitness prepare for training and avoid injury.

Most of your cardiovascular exercise should focus on running and swimming, and your strength and calisthenics training should be done to develop the necessary muscular strength and endurance for maximum pull-ups, push-ups and

sit-ups as they are necessary for success at BUD/S. Cross-training such as cycling, rowing and hiking is useful to rehabilitate an injury, to add variety or to supplement your basic training.

Work to improve your weakest areas. If you are a solid runner but a weak swimmer, don't spend all your time running just because you are good at it. Move out of your comfort zone, and spend enough time in the water to become a solid swimmer as well.

General Training Guidelines

Your workouts should be

- Planned and organized
- Gradual, steady and continual
- Consistent
- Specific
- Balanced

WEEKLY WORKOUT SUMMARY

- 1 Long Slow Distance workout for both running and swimming
- 1 Continuous High Intensity workout for both running and swimming
- 1 Interval workout for both running and swimming
- 4-5 Calisthenics Routines
- 4-6 Strength Training Sessions – 2-3 each for upper and lower body
- 4-5 Core Exercise Routines
- Daily Flexibility Routines
- Specific injury prevention exercises as needed

WORKOUTS

Long Slow Distance (LSD)

The intensity of LSD work is low to moderate, so your pace should feel relatively easy and relaxed. These workouts build endurance and provide relative recovery between more intense sessions. To determine the appropriate intensity, use the Talk Test. You should be able to talk comfortably in short sentences or phrases while training, drawing breath between phrases. If you can't speak, you are working too hard, and if you can speak continually, you are not working hard enough. For LSD workouts, focus more on duration than intensity. If you are exceptionally fit, you might perform 40-90 minutes of continuous movement in one session. A practical goal to prepare for BUD/S is to build up to comfortably running 5-6 miles or swimming 1-1.25 miles without stopping.

Continuous High Intensity (CHI)

These sessions typically involve moving for 15-20 minutes without stopping at a pace approximately 90-95% of the maximal pace you could hold for that duration. The workout should be very demanding but not totally exhausting. On a scale of 1-10, with 10 being the greatest effort possible, the workout

155

should feel like 8-9. If you are at a low fitness level, one repetition of 15-20 minutes is sufficient. As your fitness improves, 2-3 repetitions may be required. When performing more than one repetition, allow sufficient recovery between repetitions so you can maintain the desired intensity of 90-95% of maximal pace. A reasonable recovery period is approximately half of the work time. During this time, keep moving at a low intensity – slow jog, brisk walk or easy stroke. Do not come to a complete stop.

Interval (INT)

These sessions alternate short, intense work intervals with periods of recovery. The format consists of running 1/4-mile intervals or swimming 100-yard intervals, allowing a recovery period of 2-2 1/2 times the amount of time it takes to perform the work interval. Your intensity or pace should be slightly faster than the pace of your most recent 1.5-mile run or 500-yard swim. For running, your 1/4-mile interval pace should initially be about 4 seconds faster than your base pace, and for swimming, your 100-yard interval pace should initially be 2 seconds faster than your base. For example, if you recently completed a 1.5-mile run in 10:30 – 1/4 mile base pace of 1:45 – your interval training pace should be about 1:41. If you completed a 500-yard swim in 10:30 – 100-yard base pace of 2:06 – intervals should be approximately 2:04.

Begin your interval workouts with 4 intervals per session, and build progressively toward completing 10 intervals. Do not run or swim more than 10 intervals during an interval session. When you can complete 10 intervals in the prescribed times, work on gradually performing the intervals a little faster each week. Work on consistency, trying to keep little variation between your fastest and slowest interval and pacing

RUN			SWIM		
If your current pace is	Then your workout is		If your current pace is	Then your workout is	
	1/4-mile repeat time	recovery time		100-yard repeat time	recovery time
8:00-8:30	1:16-1:21	2:32-3:23	8:00-8:30	1:34-1:40	3:08-4:10
8:30-9:00	1:21-1:26	2:42-3:35	8:30-9:00	1:40-1:46	3:20-4:25
9:00-9:30	1:26-1:31	2:52-3:48	9:00-9:30	1:46-1:52	3:32-4:40
9:30-10:00	1:31-1:36	3:02-4:00	9:30-10:00	1:52-1:58	3:44-4:55
10:00-10:30	1:36-1:41	3:12-4:13	10:00-10:30	1:58-2:04	3:56-5:10
10:30-11:00	1:41-1:46	3:22-4:25	10:30-11:00	2:04-2:10	4:08-5:25
11:00-11:30	1:46-1:51	3:32-4:38	11:00-11:30	2:10-2:16	4:20-5:40
11:30-12:00	1:51-1:56	3:42-4:50	11:30-12:00	2:16-2:22	4:32-5:55
12:00-12:30	1:56-2:01	3:52-5:03	12:00-12:30	2:22-2:28	4:44-6:10
12:30-13:00	2:01-2:06	4:02-5:15	12:30-13:00	2:28-2:34	4:56-6:25
13:00-13:30	2:06-2:11	4:12-5:28	13:00-13:30	2:34-2:40	5:08-6:40
13:30-14:00	2:11-2:16	4:22-5:40	13:30-14:00	2:40-2:46	5:20-6:55
14:00-14:30	2:16-2:21	4:32-5:53	14:00-14:30	2:46-2:52	5:32-7:10
14:30-15:00	2:21-2:26	4:42-6:05	14:30-15:00	2:52-2:58	5:44-7:25
15:00-15:30	2:26-2:31	4:52-6:18	15:00-15:30	2:58-3:04	5:56-7:40
15:30-16:00	2:31-2:36	5:02-6:30	15:30-16:00	3:04-3:10	6:08-7:55

Table 1 Interval Paces

Table 1 provides appropriate paces and recovery times for interval workouts.

yourself to be fastest at the end of the workout. Every 4th or 5th week, it may be beneficial to increase your intensity using shorter, more frequent intervals. For example, 16-20 x 220-yard running intervals or 16-20 x 50-yard swimming intervals.

Allow enough recovery time to maintain the proper work intensity, without taking excessive time or wasting time. To promote faster, more complete recovery, use some active recovery, such as brisk walking, easy stroking or slow jogging for part of the time between intervals. ■

Calisthenics

During BUD/S and for the PST, you will be required to perform numerous push-ups, sit-ups and pull-ups. You should prepare specifically for these exercises. Using proper technique, perform sets of push-ups, sit-ups and pull-ups 4-5 times per week, resting 1-2 minutes between sets. Though the PST requires the exercises to be performed as rapidly as possible, you should perform most of your training exercises in a slow and controlled manner. The negative or downward portion should take at least twice as long as the positive or upward portion. Approximately once per week, perform a max set (maximal number of consecutive repetitions) to assess your progress.

Here are descriptions of each exercise as they must be performed during the PST. While training, you may occasionally do alternate versions for variety and additional fitness adaptations.

Push-up
- Begin in the up or front-leaning rest position, with feet together and palms on floor directly beneath or slightly wider than shoulders.
- Back, buttocks and legs should remain straight from head to heels at all times. Palms and toes remain in contact with the floor.
- Lower the entire body as a single unit by bending the elbows until the arms form right angles, then return to the starting position by extending the elbows, raising the body as a single unit until the arms are straight.

Variations Use caution with any push-up variation, since placing the hands in any position other than beneath the shoulders may create painful stress on the elbows.
- Include wide, narrow (triceps) and dive bomber
- Lift one foot off the floor
- Place feet on a raised surface slightly higher than the hands

Sit-up
- Begin by lying flat on floor with knees bent and heels approximately 10 inches from buttocks.
- Arms should be folded across the chest with hands touching the upper chest or shoulders. The feet may be stabilized if desired.
- Curl the body up, touching the elbows to the thighs just below the knees, keeping the hands in contact with the chest or shoulders.
- After touching elbows to thighs, lie back till the shoulder blades touch the floor.

Variations
- With fingers placed loosely behind neck (don't pull on neck), curl the trunk up and rotate so the right elbow contacts the left knee; lower trunk to floor and bring left elbow up to right knee; continue alternating rotations from right to left.
- Keeping shoulders on the floor and knees bent, alternate drawing each knee up to the opposite elbow. Return each leg so the foot rests on the floor while the other knee is drawn up.
- With arms across chest or fingers behind neck, keeping the knees bent, lift the legs

and hips off the floor drawing the knees towards the shoulders. After the abdominals have been fully contracted, lower the hips and legs until the feet touch the floor.

Note: for all abdominal exercises, keep the pelvis neutral and the lower back pressed to the floor to avoid putting stress on the lumbar spine.

Pull-up

- Begin suspended from the bar in a dead hang with arms and shoulders fully extended, palms shoulder width apart and pronated (overhand grip, facing away).
- Pull body up until chin is even with or above the top of the bar.
- Legs may be crossed or uncrossed as desired, but no kipping or jerking motions allowed.
- Lower the body in a controlled fashion until arms and shoulders are fully extended.

Variations

- Narrow or wide grip
- Supinated grip with palms toward the body to more completely isolate the biceps
- Hang from bar with hands adjacent and on opposite sides of the bar, palms facing inward in opposite directions, and alternately pull the right and left shoulders up to the bar (also called the mountain climber or commando pull-up) ▪

PUSH-UPS & SIT-UPS				PULL-UPS			
If your max is	Then your workout is			If your max is	Then your workout is		
	Sets	Reps	Total		Sets	Reps	Total
<40	5-6	10-15	50-90	<6	5-6	2-3	10-18
40-60	4-5	15-20	60-100	6-9	4-5	4-5	16-25
60-80	4-5	20-25	80-125	10-12	4-5	5-6	20-30
80-100	3-4	30-40	90-160	13-15	3-4	8-10	24-40
>100	3-4	40-50	120-200	>15	3-4	10-12	30-48

Table 2 Push-up, Sit-up, Pull-up Progression

Table 2 *provides specific training recommendations to improve your maximum number of push-ups, sit-ups and pull-ups.*

Strength Training/Weight Lifting

Muscular strength is necessary to enhance performance on the PST and increase the likelihood of success at BUD/S. It is important to gain strength properly to avoid injury.

There are many different training protocols for building strength and numerous methods of providing adequate resistance, including free weights, machines and body weight. For the purposes of this training, generally perform a single set of 8-12 repetitions (occasionally 4-6 reps or 15-20 reps) of various exercises that target major muscle groups.

You can occasionally perform a second set to provide additional training stimulus, but in most cases one set is sufficient to produce significant increases in strength. Perform a single set using a weight that cannot be lifted more than 8-12 times giving maximal effort and using proper technique. Generally perform 8-12 exercises per session.

Move from one exercise to the next quickly, only resting the amount of time it takes to set up the proper weight at the next station. This promotes overall intensity and some cardiorespiratory adaptations. Use a split routine of upper body and lower body exercises on alternate days.

To the right is a list of exercises you might incorporate into your strength program. This list is not definitive, and individuals may create personalized routines based on equipment availability and individual preferences. Alternate a variety of exercises that involve pushing (extension) with pulling (flexion) and target several major muscle groups. Avoid exercises that require high levels of skill unless you are under the supervision of a qualified coach.

Upper Body Exercises

Lat pull-downs, shoulder (military) press, biceps curl, bench press or incline press, seated row pull, deltoid lateral raise (raise arms parallel to the ground but no higher), upright row, triceps extension or dips.

Lower Body Exercises

Lunges, leg curl, back hyperextension, dead lifts, leg press or squats, and heel raises. ▨

Core Exercises

It is important to develop the strength **and endurance** of core muscles in the abdominal and spinal regions. This will improve overall body balance and alignment, improve stability and reduce injury. Sit-ups and push-ups, which should be performed regularly in preparation for BUD/S, are important core exercises. Additional core exercises include the bridge, plank, and bird dog.

Bridge

- Lie on back with knees bent and feet about ten inches from buttocks.
- Keep arms at sides or folded across the chest and keep the pelvis neutral.
- Raise the hips off the floor, creating a straight line between the knees, hips and shoulders.
- Lift the right foot off the floor and extend the leg until it is straight and creates a line from the shoulder through the hip, knee and foot.
- Meanwhile, support the body's weight by statically contracting the glutes and hamstring of the left leg. Make sure to keep the pelvis neutral and horizontal; don't let it dip toward the unsupported side.
- Hold the contraction for 3-4 seconds before lowering the pelvis to the floor with both feet near the buttocks in the original starting position.
- Lift the left foot off the floor and extend the leg while supporting the body's weight with the right leg in the same manner for 3-4 seconds.
- Continue to alternate between legs.

Plank

- Lie face down on floor with legs straight and feet together, place forearms on floor with elbows directly below shoulders, then raise body off the floor so weight is supported by toes and forearms.
- Hold body in this position by statically contracting the core muscles, maintaining a straight line from heels to shoulders.
  Variations
- Lift each arm and leg off the floor one at a time in turn, holding each position for several seconds before moving to the next position. Make sure the torso remains stable.
- Hold one arm and the opposite leg off the floor simultaneously.

Side Plank

- Lie on one side supporting body weight on one forearm with elbow below shoulder and resting the other arm along the side of the body.
- Don't let the hips sag towards the floor. Hold the spine and legs in a straight line by statically contracting the core muscles.
- Hold for desired length of time and switch to the other side.
  Variations
- Maintain core contraction while lifting the top leg off the floor by abducting the hip.
- Raise the body higher off the floor by extending the support arm completely straight and supporting the weight with one hand, meanwhile extending the opposite arm

straight above the body.

Bird Dog
- Begin on hands and knees, with hands directly below shoulders and head & neck aligned with back.
- Raise the right arm until it is fully extended and parallel to the floor. Simultaneously raise the left leg until it is fully extended. The arm, leg and back should all be in the same horizontal plane.
- Keep the torso stable; do not let the hip drop on the unsupported side.
- Hold for 3-4 seconds, then lower the upraised arm and leg to the starting position, and raise the opposite arm and leg to the same extended positions.

Superman
- Lie face down on floor with legs straight, feet together and arms straight and extended overhead.
- Keeping arms and legs straight, lift both hands and both feet several inches off the floor and hold for 3-4 seconds.
- Relax for 1-2 seconds and repeat.

Variations
- Keeping arms and legs straight, lift one hand and the opposite foot several inches off the floor and hold for 3-4 seconds. Return to starting position and simultaneously lift the other

hand and foot. Continue to alternate lifting opposite hands and feet.

Wipers
- Lie on your back with legs extended straight and together, and arms outstretched away from the body.
- Lift the legs together till they are perpendicular to the ground (hips flexed to 90 degrees). Keeping the hips flexed to 90 degrees, rotate the lower torso and pelvis to one side so the legs contact the ground.
- Rotate the lower torso and pelvis through a 180 degree arc till the legs contact the ground on the other side. Swing the legs back and forth through a 180 degree arc (like a windshield wiper). Each arc counts as one rep.
- Keep the upper back, both arms and shoulder blades in contact with the ground at all times.

Note: Effective core training is as much about learning to activate the lesser-used muscles as it is about increasing their strength. You should activate the transverse abdominis during each session. You can feel this muscle when you cough, and one technique to activate it during core exercises is to cough before performing a core exercise and to make sure you feel this muscle contracting during the exercise.

EXERCISE	WEEK				
	1-6	7-11	12-16	17-21	22-26
Bridge	2 x 20 reps (alternating)	2 x 25 reps (alternating)	3 x 20 reps (alternating)	3 x 25 reps (alternating)	3 x 30 reps (alternating)
Plank	2 x 30 sec	2 x 45 sec	3 x 40 sec	3 x 50 sec	3 x 60 sec
Side Plank (each side)	2 x 30 sec	2 x 40 sec	2 x 45 sec	2 x 50 sec	2 x 60 sec
Bird Dog	2 x 20 reps (alternating)	2 x 25 reps (alternating)	3 x 20 reps (alternating)	3 x 25 reps (alternating)	3 x 30 reps (alternating)
Superman	2 x 10 reps	3 x 8 reps	2 x 12 reps	3 x 10 reps	3 x 12 reps
Wipers	2 x 20 reps	2 x 25 reps	3 x 20 reps	3 x 25 reps	3 x 30 reps

Table 3 Core exercise Progression

***Table 3** is an example of how training might be structured. Work up to being able to complete the sets and reps listed in each time period.*

160

Flexibility requirements vary depending on the activity and the person, but you should devote some time to stretching to maintaining or enhancing flexibility. Perform stretching exercises after running and swimming workouts, while muscle and connective tissue temperature is still elevated. ▪

26-WEEK TRAINING PROGRAM

Table 4 shows how to combine all the workouts contained in this guide into a 26-week training program. This schedule of cardio and strength activities and distance targets for running and swimming over a 26 week period will help prepare you for BUD/S and the PST.

Week	MONDAY Cardio Run LSD (miles)	MONDAY Strength Upper/ Core	TUESDAY Cardio Swim CHI (min)	TUESDAY Strength Lower/Push-Sit-Pull	WEDNESDAY Cardio Run INT (reps)	WEDNESDAY Strength Core/ Push-Sit-Pull	THURSDAY Cardio Swim LSD (yards)	THURSDAY Strength Core/Push-Sit-Pull	FRIDAY Cardio Run CHi (min)	FRIDAY Strength Upper/ Core	SATURDAY Cardio Swim INT (reps)	SATURDAY Strength Lower/Push-Sit-Pull
1	3	X	15	X	4	X	1,000	X	15	X	4	X
2	3.25	X	15	X	4	X	1,100	X	15	X	4	X
3	3.5	X	16	X	5	X	1,200	X	16	X	5	X
4	3.75	X	16	X	5	X	1,300	X	16	X	5	X
5	4	X	17	X	6	X	1,400	X	17	X	6	X
6	4.25	X	17	X	6	X	1,500	X	17	X	6	X
7	4.5	X	18	X	7	X	1,600	X	18	X	7	X
8	4.75	X	18	X	7	X	1,700	X	18	X	7	X
9	5	X	19	X	8	X	1,800	X	19	X	8	X
10	5.25	X	19	X	8	X	1,900	X	19	X	8	X
11	5.5	X	20	X	9	X	2,000	X	20	X	9	X
12	5.75	X	20	X	9	X	2,100	X	20	X	9	X
13	6	X	2 x 12	X	10	X	2,200	X	2 x 12	X	10	X
14	6.25	X	2 x 12	X	10	X	2,300	X	2 x 12	X	10	X
15	6.5	X	2 x 12	X	10	X	2,400	X	2 x 12	X	10	X
16	6.75	X	2 x 14	X	10	X	2,500	X	2 x 14	X	10	X
17	7	X	2 x 14	X	10	X	2,600	X	2 x 14	X	10	X
18	7.25	X	2 x 14	X	10	X	2,700	X	2 x 14	X	10	X
19	7.5	X	2 x 16	X	10	X	2,800	X	2 x 16	X	10	X
20	7.75	X	2 x 16	X	10	X	2,900	X	2 x 16	X	10	X
21	8	X	2 x 16	X	10	X	3,000	X	2 x 16	X	10	X
22	8.25	X	2 x 18	X	10	X	3,100	X	2 x 18	X	10	X
23	8.5	X	2 x 18	X	10	X	3,200	X	2 x 18	X	10	X
24	8.75	X	2 x 18	X	10	X	3,300	X	2 x 18	X	10	X
25	9	X	2 x 20	X	10	X	3,400	X	2 x 20	X	10	X
26	9.25	X	2 x 20	X	10	X	3,500	X	2 x 20	X	10	X

Table 4 26-Week Training Program

*Perform **daily** stretching/flexibility exercises following cardio training.*

Warm-up & Cool-Down

The more intense your training session is, the longer the warm-up and cool-down periods should be. Warm-ups for LSD sessions may involve 5-10 minutes of easy jogging or paddling while gradually building the intensity to a comfortable level for beginning the workout. As the workout begins, you may continue to build intensity so that you comfortably finish the workout at a faster pace than you started. For CHI and INT workouts, you should warm up for 10-15 minutes **or more**.

Gradually build intensity from an easy jog or stroke for several minutes. Then add 4-5 high-intensity bursts lasting from 15 to 30 seconds. The warm-up should elevate your heart rate substantially, increase your breathing rate and activate a sweat response. A proper cool-down following LSD workouts may involve 2-3 minutes of easy jogging or stroking followed by 2-3 minutes of brisk walking. Time periods for CHI or INT cool-downs should be extended until you are breathing easily and your heart rate is close to its normal resting value.

BUILD YOUR OWN SCHEDULE

Weekly Schedule

Table 5 shows how a weekly workout schedule can be organized to prepare for the PST and BUD/S. An AM-PM training format such as lifting and core work in the morning and running or swimming plus stretching in the evening is best. It allows good recovery and a high quality of work for each session. However, if necessary, all training can be performed in one extended block of time. If performing several activities in one session, perform your weakest activity first while you are still fresh. Avoid over-exercising a body part with too many exercises or activities in the same day. Note that the schedule does not place upper body strength training and swimming or lower body strength training and running on the same days.

Since there is some overlap between the demands of weight lifting, calisthenics and core exercises, do not combine more than two of these routines on a given day.

You can do some calisthenics and core training on the same day as strength training, but don't exhaust yourself with all routines on the same day. If you are already doing higher LSD mileage, you may begin at a later week in the program or add a second LSD session (see **Table 7**). You should always begin CHI and INT portions of the program at Week 1.

	MONDAY	TUESDAY	WEDNESDAY	THURSDAY	FRIDAY	SATURDAY
Run	LSD		INT		CHI	
Swim		CHI		LSD		INT
Lift	Upper	Lower			Upper	Lower
Calisthenics		X	X	X		X
Core	X		X	X	X	
Flexibility	X	X	X	X	X	X

Table 5 Weekly Training Schedule

Progression

Gradually build up your workload from a safe, manageable level to the highest level of fitness possible in the time you have available before you take the PST or attend BUD/S.

Table 6 shows how to increase your workload across the different training bands over 26 weeks. If you are at a high level of fitness, you may choose to begin with a higher training volume such as a 5-mile run (as indicated in Week 9) rather than a 3-mile run.

Week	LSD		CHI	INT
	Run (miles)	Swim (yards)	Run/Swim (mintutes)	Run/Swim (reps)
0	1.5 (timed)	500 (timed)		
1	3	1,000	15	4
2	3.25	1,100	15	4
3	3.5	1,200	16	5
4	3.75	1,300	16	5
5	4	1,400	17	6
6	4.25	1,500	17	6
7	4.5	1,600	18	7
8	4.75	1,700	18	7
9	5	1,800	19	8
10	5.2	1,900	19	8
11	5.5	2,000	20	9
12	5.75	2,100	20	9
13	6	2,200	2 x 12	10
14	6.25	2,300	2 x 12	10
15	6.5	2,400	2 x 12	10
16	6.75	2,500	2 x 14	10
17	7	2,600	2 x 14	10
18	7.25	2,700	2 x 14	10
19	7.5	2,800	2 x 16	10
20	7.75	2,900	2 x 16	10
21	8	3,000	2 x 16	10
22	8.25	3,100	2 x 18	10
23	8.5	3,200	2 x 18	10
24	8.75	3,300	2 x 18	10
25	9	3,400	2 x 20	10
26	9.25	3,500	2 x 20	10

Table 6 Workout Progression

More Time to Prepare

Beyond 26 weeks, do not increase INT or CHI distances. Rather, focus on gradually and progressively increasing intensity for the set distances of these workouts. You can also increase your LSD work by performing longer sessions and/or increasing the number of sessions per week as shown in **Table 7**. However, beyond 9-10 miles of running per week and 3,500-4,000 yards of swimming per week, the improvements in fitness become proportionately smaller relative to the time invested. If you perform large amounts of LSD work, be sure to keep the pace relatively relaxed.

As your fitness improves, occasionally incorporate a longer session of activity (2-3 hours) such as hiking, canoeing, road cycling or mountain biking at a comfortable but steady pace to improve physical and mental endurance. Continue to progressively increase your muscular strength and endurance using the calisthenics, strength and core routines already established. ■

	MONDAY	TUESDAY	WEDNESDAY	THURSDAY	FRIDAY	SATURDAY
Run	LSD 8 miles		INT 10 x 1/4 mile	LSD 4 miles	CHI 2 x 20 minutes	
Swim	LSD 1,500 yards	CHI 2 x 20 minutes		LSD 3,000 yards		INT 10 x 100 yards
Lift	Upper	Lower			Upper	Lower
Calisthenics		X	X	X		X
Core	X			X	X	X
Flexibility	X	X	X	X	X	X

Table 7 Weekly Training Schedule (Increased LSD Sessions)

Strong in one thing; weak in another

If you have unbalanced fitness – you are clearly slower in either running or swimming – you should devote a greater percentage of your training to improve the slower activity. SEAL candidates with a swim time slower than 10:35 or a run time slower than 10:38, while performing moderately or well in the other activity, should focus more attention on the slower event. **Table 8** is an example of a schedule weighted toward improving a slower swimmer. A strong swimmer with limited running ability would reverse the schedule. ■

	MONDAY	TUESDAY	WEDNESDAY	THURSDAY	FRIDAY	SATURDAY
Run		INT			LSD	
Swim	LSD		CHI	LSD		INT
Lift		Upper	Lower		Upper	Lower
Calisthenics	X		X	X		X
Core	X	X		X	X	
Flexibility	X	X	X	X	X	X

Table 8 Weekly Training Schedule For A Slow Swimmer

Keep a record of your training. You will see your progress and have a history to show to a mentor or coach. A tangible record of your performances allows you to establish specific goals and can increase your motivation to train. Training records make it easier to avoid training mistakes or recognize potential problems before they become serious. Record basic information such as time and distance for running and swimming workouts (including individual times for each interval during interval workouts); number of reps of calisthenics and core exercises; and details of strength workouts (exercises, sets, reps, and amount of weight lifted). You may also choose to record more detailed information such as notes about your diet, the environment (temperature, humidity, wind), psychological state of mind (relaxed, anxious, energized, listless), amount of sleep, persistent soreness or any other variable that might affect your training. ■

Available at www.sealswcc.com/pdf/naval-special-warfare-physical-training-guide.pdf

Naval Special Warfare BUD/S Injury Prevention

Prevention is the key to avoiding a visit to BUD/s medical for a movement related injury. It is too late to begin a preventative program once symptoms of an injury have been identified. Therefore, it is vital that BUD/s candidates properly prepare themselves to complete the rigorous training injury free. This preparation includes stretching and strengthening of specific muscles not usually addressed in a typical BUD/s candidate workout routine.

UPPER EXTREMITY INJURY PREVENTION

Stretching Exercises

- Hold each for 30 seconds
- Repeat 3 times
- Perform 2 times a day

Stretches should NOT be painful!

Upper extremity injuries commonly encountered in BUD/s usually involve the shoulder, and are easily prevented with a consistent stretching and strengthening program.

- Scalene Stretch
- Lat and Prayer Stretch
- Pectoral Stretch
- Thumb-Down Shoulder Strength
- Lawnmower Pull (Mid Back)

- Shoulder External Rotation
- Shoulder Internal Rotation
- The "Y" Exercise
- The "T" Exercise
- The "W" Exercise

■

Scalene Stretch

The neck muscles are instantly put to the test when candidates are asked to carry boats on their heads. Nerve injuries often originate from the neck, and this stretch provides flexibility to these muscles to prevent nerve compression.

Lat and Prayer Stretch

Tight upper back muscles prevent students from maximizing their overhead reach. This can lead to shoulder injuries during log PT. These stretches can be performed sitting with elbows on an elevated table (left) or kneeling (right). The stretch should be felt on the outside of the armpit and into the back.

Pectoral Stretch

Due to the high volume of push-ups, pectoral muscles become strong and tight limiting shoulder range of motion and making the joint vulnerable to injury. This stretch should be performed in a corner so both shoulders can be stretched. Stretch should be felt in the chest, and arms can be moved up and down to change where the stretch is felt.

Thumbs-Down Shoulder Strength

Perform exercise with a rubber band or a cable machine. Arm should be oriented 45 degrees from midline of the body with thumb pointing down. Raise arm to shoulder height. This exercise will strengthen a commonly injured muscle on the top portion of the shoulder and should not be painful!

Lawnmower Pull (Mid Back)

Perform exercise with a rubber band or a dumbbell. One arm at a time, and stagger your feet for balance. Lean forward with your back straight, and with a rowing motion, pull weight back towards your side. Make sure to cover the full range of motion, and your shoulder blade should move toward the center of your back.

Shoulder External Rotation

Perform this exercise with rubber bands or a cable machine. Start with your hand grasping handle across your body with your elbow firmly against your side. Pull the cable away from your body while maintaining contact between your elbow and your side. Use a rolled up towel to hold between elbow and side. If the towel falls while performing this exercise, your elbow is leaving your side, and you are performing this exercise incorrectly.

BUD/S INJURY PREVENTION PAGE 5
WWW.SEALSWCC.COM

Shoulder Internal Rotation

Perform this exercise with rubber bands or a cable machine. Start with your hand away from your body and elbow firmly against your side. Pull weight towards body, stopping once your hand is directly in front of you. Keep your elbow against your side throughout the exercise. Again, use a towel against side if needed.

The "Y" Exercise

This exercise should be performed on your stomach while lying on a table or exercise ball. Start with arms hanging in a Y position below your body, and finish with them held in the air while maintaining the Y position. Elbows should be kept straight, and thumbs should point up. This exercise should be felt in the lower trap, or middle of your back.

Available at www.sealswcc.com/PDF/naval-special-warfare-injury-prevention-guide.pdf

INDEX